Calculus for Cranks

Calculus for Cranks

by Nets Hawk Katz

Yale

UNIVERSITY PRESS

New Haven and London

Yale University Press books may be purchased in quantity for educational, business, or promotional use. For information, please e-mail sales.press@yale.edu (U.S. office) or sales@yaleup.co.uk (U.K. office).

Printed in the United States of America

Library of Congress Control Number: 2020937478

ISBN 978-0-300-24279-9 (pbk. : alk. paper)

A catalogue record for this book is available from the British Library.

This paper meets the requirements of ANSI/NISO Z39.48-1992 (Permanence of Paper).

10 9 8 7 6 5 4 3 2 1

Contents

PREFACE

This book began as class notes for the Caltech course Math 1a which I taught in the fall of the years 2013, 2014, 2015, and 2016. The California Institute of Technology is an unusual institution. All students must have a major (we call them options) in the sciences or in engineering. These are declared in spring of the first year. But we also have a serious core curriculum, representing material that we believe all scientists should know. Sometimes the core courses are split into "tracks." For instance, 1b and 1c have "practical" and "analytic" tracks, whose names don't reflect well what they are but do reflect that different students have different levels of tolerance for rigor and abstraction in their mathematics curriculum. But Math 1a has no tracks and is intended as a common experience for all entering Caltech undergraduates. It is called simply Calculus, but it is a rigorous foundation for all of single variable calculus. Such courses exist elsewhere as Honors Calculus for early math majors and as terminal courses in real analysis or advanced calculus for upper-level math majors. But currently, Caltech is one of very few places in the United States where this material considered suitable or important for a typical science major. Even the Greek letters ϵ and δ, connoting smallness which make up the core of the course, are considered too frightening for any but a mathematician to read. But at Caltech, we believe this material is accessible and important for all scientists. How can we be right and the rest of the world wrong? It is a question often asked by visitors to our department.

The title of the book came to me at an earlier stage in my career. I was a professor at Indiana University Bloomington between 2004 and 2012. I taught a variety of classes aimed at different audiences. I rarely found it fully satisfying. Too often, students viewed their courses as hoops to be jumped through, not as arrays of theories and views of the world to be learned. A student interested in the sciences or in theoretical economics would take an "early transcendals" course using Stewart's "Calculus." (This was a class I taught.) A student interested in business might take Business Calculus, which excluded trigonometric functions (but not exponentials) and generally encouraged a lower level of independence. A student of the humanities would avoid all this and take a course called Finite Math. It was a rare student indeed for which any of these was a life-changing experience. And the faculty, complying, rarely invested too much of themselves into these canned courses that kept the gates for standard career paths.

One very nice feature of Indiana University was the existence of the Collins Living Learning Center. Locals referred to it as the Collins College. This Bohemian institution existed primarily for students in the arts and humanities who wanted an unusual level of control over their education. The Collins College solicited proposed courses from members of the university community. In the end, these proposals were voted on by the students of the college. Around 2010, I proposed a course entitled "Calculus for Cranks." The cranks in question were the Collins College students.

The word "cranks" has an extremely negative connotation within mathematics. It usually refers to people who are doing math wrong while being sure they are right. (The interested reader should consult Underwood Dudley's "Mathematical Cranks" which is the definitive treatise on the subject.) But I wanted to fight the establishment. My cranks would do mathematics correctly when the world is telling them they shouldn't. I discussed this with some students in the Collins College who were scouting for courses. They were enthusiastic. The next step was to give a sample lecture to the Collins College student body as a whole. My lecture was about the real numbers. I explained that real numbers are infinite decimal expansions. I reminded them that in elementary school, they were taught that to add and multiply numbers, you start at the right. But for an infinite decimal expansion, you can never get to the right. They had been lied to in the seventh grade when real numbers and their operations were introduced without a definition. The only way to sort it out was to take some kind of a limit. They had been doing calculus or analysis, all along, without knowing it. I would reveal the machine behind the curtain. At the end of the lecture, a Collins College student who happened to be a math major got up to denounce me. She did not understand what I was saying and was therefore certain the typical Collins College student who had no need of math would not understand me either. The Collins College students went through a democratic process to decide this issue. I lost. They decided that they were not cranks.

This incident stuck in my craw. When I had the opportunity to move to Caltech and learned about Math 1a, I jumped at it because it proved that Caltech students are all cranks. I insisted on being allowed to teach the course at the first possible opportunity and I insisted on writing my own notes. Caltech students, being all cranks, want, like pioneers, to understand every part of science and math to its foundations. That was what I wanted to give them in Math 1a.

Does it make sense? Why should anyone be taught the real numbers? It is precisely because they are an abstraction. Their name is propaganda of the highest order. They are not, in fact, real. When a scientist collects data, the scientist is getting an approximation with a level of accuracy that is somewhat arbitrary; that is, it depends on the quality of the scientist's equipment. The real numbers are the right thing to study because they are exactly the abstraction that lets you study approximations with arbitrary levels of accuracy. To see things that way, you have to look inside the proof. Your goal isn't to show the function is continuous or differentiable. Your goal is to find out how δ depends on ϵ. Your goal isn't to know the statement of a theorem which may have seemed obvious to you for a long time. Your goal is to unlock the algorithm that lives inside the proof. Become enough of a crank, and even if you're not a mathematician but a scientist or engineer in a math-heavy discipline, you can rise above your masters.

If you read this text and are familiar with earlier attempts at rigorous calculus texts like those by Apostol, Kitchen, Spivak, and others, you may find the current approach a bit different. It is specifically aimed at the cranks of Caltech. Real numbers are described as decimal expansions to better link the material with the math, such as it is, that is typically presented in high school. Less emphasis is placed on theorems that can be used repeatedly, such as L'Hopital's rule, and more on techniques that can be modified easily for different situations.

Cranks can exist at places other than Caltech, and this material is aimed at them too. They should be aware that their counterparts here have had a high school education with at least a BC level calculus course, which is equivalent to two semesters of calculus at most colleges and universities. However, like anyone who took a low-level mathematics course, they may not have an entirely clear understanding of it. The time to start understanding is now. Don't let people tell you that understanding isn't for you.

Nets Hawk Katz
IBM Professor of Mathematics
Caltech

Pasadena, California
August, 2020

Chapter 1

INDUCTION AND THE REAL NUMBERS

◇ 1.1 Induction

This course, Calculus for Cranks, will be somewhat unusual. It is a proof-based treatment of calculus, intended for those who have already demonstrated a strong grounding in the calculus, as treated in a high school course. You might ask, "What's the point?" "Will I learn anything new?" "Will it be at all useful for applied work?" "Are formal proofs just voodoo that has no impact on 'the right answers?'" Mathematicians usually defend a course like this in a philosophical vein. We are teaching you how to think and the ability to think precisely and rigorously is valuable in whatever field you pursue. There is some truth in this idea, but we must be humble in its application and admit that the value of being able to think does depend somewhat on what is being thought about. You will be learning to think about analysis, the theoretical underpinning of calculus. Is that worth thinking about?

A fair description of the way I hope most of you already understand calculus is that you are familiar with some of its theorems (the fundamental theorem, and rules for differentiation and integration) and you know some ways of applying them to practical problems. Why then study their proofs? Different answers are possible. If your interest is in applying calculus to the real world, the proofs of the theorems have surprisingly much to say about the matter. If you are a scientist or an engineer, usually data in the real world come to you with limited measurement accuracy, not as real numbers but as numbers given to a few decimal places with implicit error intervals. Never-

theless, studying the abstraction of the real numbers, as we
shall do in later sections, tells us what we know reliably about
the way in which these errors propagate. (Indeed all of analysis concerns the estimation of errors.) Another subtler reason
for the study of theory is that there is more to calculus than
strictly the statements of the theorems. The ideas of calculus can be recycled in slightly unfamiliar settings, and if you
don't understand the theory, you won't recognize these ideas.
We will start to see this immediately, in discussing the natural
numbers, and it will be a recurring theme in the course.

For our purposes, the natural numbers are the positive integers. They are denoted as **N**, the set of natural numbers.

$$\mathbf{N} = \{1,2,\ldots,n,\ldots\}.$$

Here on the right side of the equation, the braces indicate that
what is being denoted is a set (a collection of objects). Inside
the braces, are the objects in the set. The ... indicate that I
am too lazy to write out all of the objects (there are after all
infinitely many) and mean that I expect you to guess based on
the examples I've put in $(1,2,n)$ what the rest of the objects
are. An informal description of the natural numbers is that
they are all the numbers you can get to by counting, starting
at 1.

Most of you have been studying the natural numbers for at
least the last thirteen years of your life, based on some variant
of the informal description. Indeed, most of what you have
learned about the natural numbers during your schooling has
been true. Mathematicians can be expected to be unsatisfied
with such descriptions, however, and to fetishize the process of
writing down a system of axioms describing all the properties
of the natural numbers. There is such a system, called the
Peano axioms, but I will dispense with listing them except for
the last, which details an important method of proof involving
the natural numbers, that we will use freely.

The principle of induction

Let $\{P(n)\}$ be a sequence of statements indexed by the natural numbers. (The fact that I denote the n dependence of the statement $P(n)$ indicates that it is a member of a sequence.) Suppose that $P(1)$ is true and suppose that if $P(n)$ is true, it follows that $P(n+1)$ is true. Then $P(n)$ is true for all natural numbers n.

When carrying out a proof using the principle of induction, proving $P(1)$ is called the *base case*. Proving that $P(n)$ implies $P(n+1)$ is called the *induction step*. The given statement $P(n)$ used to imply $P(n+1)$ is called the *induction hypothesis*.

In case this statement of the principle of induction is too abstract, I will give a number of examples in this section, indicating how it can be used. I begin by saying, however, that the principle of induction is very closely related to the informal description of the natural numbers, namely that all natural numbers can be reached from 1 by counting. Here is an informal proof of the informal description by induction. Don't take this too seriously if you prefer that all your terms be defined.

Proof of the informal description of the natural numbers

Let $P(n)$ be the statement "n can be reached from 1 by counting." Clearly 1 can be reached from 1 by counting. So $P(1)$ is true. Suppose $P(n)$ is true. Then n can be reached from 1 by counting. To reach $n+1$ from n by counting, just do whatever you did to reach n by counting and then say "$n+1$." Thus $P(n)$ implies $P(n+1)$. Thus the principle of induction says that $P(n)$ is true for all n. Thus we know that every natural number n can be reached from 1 by counting.

The most important statement that we will prove in this section using induction is the principle of well ordering. We will use well ordering when understanding the real number system. Instead of setting up the real numbers axiomatically, I will describe them as they have always been described to you, as infinite decimal expansions. I will use the well ordering principle to obtain an important completeness property of the reals: the least upper bound property.

Theorem 1.1.1 Well ordering principle

Every nonempty set of natural numbers has a smallest element.

Proof of Well ordering principle

We will prove this by proving the contrapositive: Any set A of natural numbers without a smallest element is empty. Here's the proof: Let A be a set of natural numbers without a smallest element. Let $P(n)$ be the statement: Every natural number less than or equal to n is not an element of A. Clearly $P(1)$ is true, because if 1 were an element of A, it would be the smallest element. Suppose $P(n)$ is true. Then if $n + 1$ were an element of A, it would be the smallest. So we have shown that $P(n)$ implies $P(n+1)$. Thus by induction, all $P(n)$ are true so no natural number is in A. Thus A is empty.

What is the value of a proof? Often a proof consists of an algorithm that we could implement as programmers. Suppose we're presented with a set of natural numbers and a way of testing whether each natural number belongs to the set. To find the smallest element, we count through the natural numbers, checking each one in turn to see if it belongs to the set. If the set is nonempty, we are guaranteed that this algorithm will terminate. That is the practical meaning of the above proof.

A common example to demonstrate proof by induction is the study of formulas for calculating sums of finite series. An example with a rich history is

$$S_1(n) = 1 + 2 + 3 + \cdots + n = \sum_{j=1}^{n} j.$$

(Here I wrote the sum first with . . . , assuming you knew what I meant [the sum of the first n natural numbers] and then wrote it in summation notation which is somewhat more precise.) Legend has it that when the great mathematician Gauss was in grade school, his teacher asked the whole class to compute $S_1(100)$, hoping to take a coffee break. Before the teacher left the room, Gauss yelled out 5050. How did he do it? He first wrote the sum forwards,

$$1 + 2 + \cdots + 100.$$

then backwards,

$$100 + 99 + 98 + \cdots + 1.$$

Then he added the two sums vertically getting 101 in each column. Thus, twice the sum is 10100. So the sum is 5050.

Applying Gauss's idea to a general n, we get

$$S_1(n) = \frac{n(n+1)}{2} = \binom{n+1}{2}.$$

A common example of a proof by induction is to prove this formula for $S_1(n)$. We dutifully check that $\binom{1+1}{2} = 1$, verifying the formula for $n = 1$. We assume that $\binom{n+1}{2}$ is the sum of the first n natural numbers. Then we do a little algebra to verify that $\binom{n+2}{2} - \binom{n+1}{2} = n + 1$, concluding that $\binom{n+2}{2}$ is the sum of the first $n + 1$ natural numbers. We have thus shown by induction that the formula is true for all n.

Gauss's proof seems like a lot more fun. The proof tells us the answer, finding the formula for the sum. The induction proof seems like mumbo jumbo certifying the formula after we already know what it is.

We can formalize Gauss's approach set-theoretically (or, depending on your point of view, geometrically or combinatorially) by viewing the goal of Gauss's sum: to count the number of elements in a set. We let

$$A = \{(j,k), j,k \in \mathbf{N}^2 : 1 \leq j \leq k \leq n\}.$$

Then the number of elements of A, which we denote as $|A|$ is exactly $S_1(n)$. That is because all the elements of the set A with second component 1 are (1,1). All elements with second component 2 are (1,2) and (2,2) and so on. The jth column has j elements. So when we count all the elements, we are adding the numbers j for all j between 1 and n. In this context, Gauss's idea is to switch the role of rows and columns.

Let

$$B = \{(j,k), j,k \in \mathbf{N}^2 : 1 \le k \le j \le n\}.$$

Clearly

$$|A| = |B|.$$

Moreover, every point (j,k) with j,k natural numbers less than or equal to n is either in A, or B, or both. This is because either $j \le k$ or $k \le j$, with both true when $j = k$. We have

$$|A \cup B| = n^2,$$

since there are n^2 points in the $n \times n$ square. But

$$|A| + |B| \ne |A \cup B|,$$

because the left-hand side double-counts the points that are in both. In fact,

$$|A| + |B| = |A \cup B| + |A \cap B| = n^2 + n.$$

Noting that the left-hand side is $2S(1)$, we have reproduced Gauss's result.

Before leaving Gauss's proof, let us at least examine how it generalizes to sums of squares. Let us consider

$$S_2(n) = 1 + 4 + \cdots + n^2 = \sum_{j=1}^{n} j^2.$$

In order to calculate this sum, à la Gauss, it helps to have a geometric notion of the number j^2. It is in fact the number of pairs of natural numbers less than or equal to j. In set-theoretic notation

$$j^2 = \#\{(l,m); l,m \in \mathbf{N}, l,m \le j\}.$$

Thus we can write $S_2(n)$ as the number of elements of a set of triples. Basically we use the third component of the triple to write down which term of the sum the ordered triple belongs to.

$$S_2(n) = |\{(j,l,m) : j,l,m \in \mathbf{N}, l,m \leq j, j \leq n\}|.$$

Thus the number we seek, $S_2(n)$, is the number of triples of natural numbers less than or equal to n so that the first component is greater than or equal to the last two components. Gauss's trick generalizes to the following observation. For any ordered triple, one of the components is at least as big as the other two. This suggests we should compare three copies of $S_2(n)$ to n^3, which is the number of triples of natural numbers less than or equal to n. But we have to be careful: we are counting triples where there are two components larger than the third twice and we are counting triples where all three components are equal three times.

Now observe that the number of triples of natural numbers less than or equal to n with all components equal, formally

$$\#\{(j,j,j) : j \in \mathbf{N}, j \leq n\},$$

is just equal to n, the number of choices for j. It is also easy to count triples that have the first two components large and the third smaller. We observe that

$$\#\{(j,j,l) : j,l \in \mathbf{N}, j \leq n, l \leq j\} = S_1(n).$$

We get this because for each j there are j choices of l, so we are summing the first n numbers. Then like Gauss we observe that each triple with two equal components at least as large as the third, has the third component somewhere. Combining all these observations, we can conclude that

$$n^3 = 3S_2(n) - 3S_1(n) + n.$$

(Basically the first term correctly counts triples with all different components. The first term double-counts triples with two equal components and one unequal components but the second term subtracts one copy of each of these. The first term triple-counts triples with all components the same, but the second term also triple counts them, so we have to add n to correctly account for all triples.) Since we already know the formula for $S_1(n)$, we can solve for $S_2(n)$ and a little algebra gives us the famous formula,

$$S_2(n) = \frac{n(2n+1)(n+1)}{6},$$

which you may have seen before.

As you can imagine, the argument generalizes to the sum of kth powers.

$$S_k(n) = \sum_{j=1}^{n} j^k.$$

Keeping track of $k+1$-tuples with some entries the same is tricky, but the highest-order term in the formula is easy to guess. Gauss's trick is that $k+1$-tuples with a single largest component have that component in one of $k+1$ places. So what we get is that

$$S_k(n) = \frac{1}{k+1} n^{k+1} + lower\ order\ terms.$$

You know some calculus so this should be familiar to you as one of the most basic facts in calculus. It encodes that the indefinite integral of x^k is $\frac{1}{k+1} x^{k+1}$. That's the same factor of $\frac{1}{k+1}$ in both places. So what's actually happening is that Gauss's trick gives you a new (and perhaps unfamiliar) way of proving this fundamental fact. In your previous study of calculus, how might you have derived it? Basically you work through the fundamental theorem of calculus , you know how to take derivatives and you know you have to guess a function whose derivative is x^k. (Ask yourself: how are these different proofs related?)

How does induction fit in? Let me ask an even more general question. Pick some function f acting on the natural numbers. Define the sum

$$S_f(n) = \sum_{j=1}^{n} f(j).$$

[To those already familiar with calculus, this sum is in analogy to the integral of f.] Now, we can only calculate this by induction if we can guess an answer. Let's consider a guess

$F(n)$ for this sum. What has to be true for induction to confirm that indeed $S_f(n) = F(n)$. First we have to check that $S_f(1) = f(1) = F(1)$. Otherwise, the formula will already be wrong at $n = 1$. Then we have to check that

$$F(n + 1) - F(n) = S_f(n + 1) - S_f(n) = f(n + 1),$$

concluding that the formula being correct at n implies that the formula is correct at $n + 1$. If you stare at this for a moment, you'll see that this is in direct analogy to the fundamental theorem of calculus. The difference $F(n+1) - F(n)$ plays the role of the derivative of F. The sum plays the role of the integral of f. To calculate the sum, you have to guess the antiderivative. Induction here plays the role of the calculus you already know and the unfun guess is a process you're familiar with.

To sum up: we've learned in this section about proofs by induction. We used induction to prove the well ordering principle. In calculating finite sums, induction plays the same role as the fundamental theorem of calculus. I hope I'm also starting to convince you that proofs have meaning and that we can learn surprising and interesting things by examining their meaning. If you starting taking this section overly seriously, you might conclude that the calculus you know doesn't need the real numbers in order to operate. It mostly consists of algebraic processes that work on the natural numbers as well. That isn't what this course is about, however. Soon we'll begin studying the real numbers and much of the focus of this course will be on what is special about them.

Exercises for Section 1.1

---

1. Let $p(x)$ be a polynomial of degree n with integer coefficients. That is,

$$p(x) = a_n x^n + a_{n-1} x^{n-1} + \cdots + a_0,$$

where the coefficients $a_0, \ldots, a_n$ are integers and where the leading coefficient a_n is nonzero. Let b be an integer. Then show that there is a polynomial $q(x)$ of degree $n-1$ with integer coefficients and an integer r so that

$$p(x) = (x - b)q(x) + r.$$

[In other words, you are asked to show that you can divide the polynomial $p(x)$ by the polynomial $(x - b)$ and obtain an integer remainder.] (Hint: Use induction on n. To carry out the induction step, see that you can eliminate the leading term, and then use the induction hypothesis to divide a polynomial of degree $n - 1$.)

2. Use Gauss's trick to find a formula for the sum of the first n fourth powers. To verify your calculation, prove by induction that this formula is correct.

3. Let $S_k(n)$ denote the sum of the first n kth powers. Prove by induction that $S_k(n)$ is a polynomial (whose coefficients are rational numbers) of degree $k + 1$ in n. (Hint: You should prove this by induction on k. You should use as your induction hypothesis that $S_j(n)$ is a polynomial of degree $j + 1$ for all j smaller than k. [This is sometimes called strong induction.] The section should give you a good guess for what the leading term of $S_k(n)$ should be. Express the rest of $S_k(n)$ as a combination of $S_j(n)$'s for smaller j.)

4. Prove the principle of strong induction from the principle of induction. That is let $Q(n)$ be a sequence of statements indexed by the natural numbers. Suppose that $Q(1)$ is true. Moreover, suppose that the first n statements $Q(1), Q(2), \ldots, Q(n)$ together imply $Q(n + 1)$. Then $Q(n)$ is true for all natural numbers n. (Hint: Define statements $P(n)$ to which you can apply the principle of induction.)

5. As in the text, define the bino-
mial coefficient $\binom{k}{2} = \frac{k(k-1)}{2}$
and the binomial coefficient
$\binom{k}{3} = \frac{k(k-1)(k-2)}{6}$. These represent respectively, the num-
ber of ways of choosing two
natural numbers from the first
k, and the number of ways of
choosing three natural numbers from the first k. Find a
formula for the sum $\sum_{k=1}^{n} \binom{k}{2}$.
Check your formula using in-
duction. (Hint: Observe that
a choice of three elements
from the first k can be bro-
ken into two parts.   First
you choose the smallest of the
three and then you choose the
other two. Compare this description to the sum.)

6. Prove the identity $\binom{n}{k}$ =
$\binom{n-1}{k-1} + \binom{n-1}{k}$. (Hint: write
both terms on the right hand
side with the common denominator $k!(n-k)!$.)   Use the
identity you just proved to
prove by induction the identity

$$\sum_{j=0}^{n} \binom{n}{j} = 2^n.$$

7. Prove by induction on k that
the sum

$$S_k(n) = \sum_{j=1}^{n} j^k,$$

is a polynomial of degree $k+1$
with leading term $\frac{1}{k+1}n^{k+1}$.
(Hint: Observe by telescoping
that

$$\frac{1}{k+1}n^{k+1}$$

$$= \sum_{j=1}^{n} \frac{1}{k+1}(j^{k+1}-(j-1)^{k+1}).$$

Use the induction hypothesis
to control the error terms.)

Flipped recitation, induction

In the section on induction, you learned the method of proof called the principle of induction. We saw how to use it to find exact formulas for certain sums.

Problem 1 Telescoping sums

Let f be a rational-number-valued function defined on all the natural numbers. Consider

$$S(n) = \sum_{j=1}^{n} f(j+1) - f(j).$$

Show by induction that

$$S(n) = f(n+1) - f(1).$$

Problem 2

Use the result of problem 1 to calculate

$$S(n) = \sum_{j=1}^{n} \frac{1}{j(j+1)}.$$

Hint: Write $\frac{1}{j(j+1)}$ as the difference between two fractions.

Problem 3

Use the result of problem 1 to calculate

$$S(n) = \sum_{j=1}^{n} 4j^3 + 6j^2 + 4j + 1.$$

Hint: Write the summand as $(j+1)^4 - j^4$.

◇ 1.2   Induction for derivatives of polynomials

One of the central facts from a typical course in high school calculus is that the derivative $f'(x)$ of the function $f(x) = x^k$ is given by

$$f'(x) = kx^{k-1}.$$

For us, it seems premature to be talking about this. We have not yet defined limits or derivatives, which we'd like to understand for a considerably more general class of functions than polynomials. But in this section, we're going to do something a bit odd. We will restrict our attention to polynomials, and we will use the high school formula as the definition of the derivative. And then, we will use induction to discover a number of fundamental facts about polynomials. This serves different purposes. It gives a rich set of problems on which we can bring induction to bear and it allows us to emphasize certain facts about polynomials which will play an important role later.

First, we should define a polynomial. We will for the moment restrict the coefficients of my polynomials to be rational numbers although this restriction is not essential. We're just trying to stay in the spirit of things. We consider rational numbers to be simple objects which we understand, but shortly we will undertake rather exacting work to make sense of what real numbers are. After we have done this, there will be no reason to restrict ourselves to polynomials with rational coefficients.

polynomial of degree k   Let $a_0, a_1, \ldots, a_k$ be rational numbers and x be a variable. Then an expression of the form

$$a_0 + a_1 x + \cdots + a_k x^k$$

will be called a polynomial (of degree k).

At this moment, we won't consider polynomials as functions at all. They are merely expressions that look like functions. The ingredients to make a polynomial are its coefficients, which in the definition we label $a_0, a_1, \ldots, a_k$ and its variable which we label x. To identify and work with a polynomial, it will be important to be clear on which are the coefficients and which is the variable. Always ask yourself , when we present a polynomial, what is the variable. Some of the fancy footwork which will follow will require careful choices of the variable.

derivative of a polynomial Let $p(x) = a_0 + a_1 x + \cdots + a_k x^k$ be a polynomial. We define its derivative, the polynomial $p'(x)$ by

$$p'(x) = a_1 + 2a_2 x + \cdots + k a_k x^{k-1}.$$

All we have done here is to take the high school formula as a definition. But when we do this, we have no idea whether our definition is natural or makes any sense. We don't know if the derivative follows nice rules. To be able to make good use of our definition, it will help to establish the product rule and the chain rule. Doing this is a little tedious. We will do it here in special cases and leave the general case as an exercise. In order for any of this to make sense, we should define the sums, products and compositions of polynomials.

Combinations of polynomials Let $p(x) = a_0 + a_1 x + \cdots + a_k x^k$ and $q(x) = b_0 + b_1 x + \cdots + b_l x^l$ be polynomials. The sum $p(x) + q(x)$ is the polynomial whose jth coefficient c_j is given by $a_j + b_j$.

multiplication Here we consider the coefficients a_j and b_j to be zero, when j is larger than the degree of p or q respectively. Then $p(x)q(x)$ is the polynomial whose jth coefficient c_j is given by

$$c_j = \sum_{m=0}^{j} a_m b_{j-m}.$$

(This formula just combines all terms in the expansion of the product of the two polynomials which contain an x^j.)

composition   The composition $p(q(x))$ is given by

$$p(q(x)) = a_0 + a_1 q(x) + \cdots + a_k (q(x))^k.$$

(Here the meaning of the powers of q comes from the previous definition of multiplication of polynomials.)

Theorem 1.2.1
Product rule for monomials

Let $p(x) = x^k$ and $q(x) = x^l$. Then with

$$r(x) = p(x)q(x),$$

we have that

$$r'(x) = p(x)q'(x) + p'(x)q(x).$$

Proof of the product rule for monomials

The polynomial $r(x)$ is of course x^{k+l}. By definition we have

$$r'(x) = (k+l)x^{k+l-1}$$

$$= lx^{k+l-1} + kx^{k+l-1}$$

$$= p(x)q'(x) + p'(x)q(x).$$

We'll leave the proof of the product rule for polynomials in the general case as an exercise. This might seem much harder, but it is actually just a matter of breaking up the product as a sum of products of individual terms. For now, we'll accept the product rule as known:

Theorem 1.2.2 Product rule for polynomials
Let $p(x)$ and $q(x)$ be polynomials and let $r(x) = p(x)q(x)$ be their product then

$$r'(x) = p(x)q'(x) + p'(x)q(x).$$

Now, we will establish the chain rule in the case where the first polynomial is a monomial.

Theorem 1.2.3 Chain rule for monomials
Let $q(x)$ be a polynomial. (You were expecting me to say that $p(x)$ is also a polynomial, but here $p(x)$ will be x^n.) Let $r(x) = (q(x))^n$. Then

$$r'(x) = n(q(x))^{n-1}q'(x).$$

Proof of the chain rule for monomials
This is clearly a job for induction. When $n = 1$, we are just differentiating $q(x)$ and we are getting $q'(x)$ as the derivative. So the base case checks out. Now suppose we already know how to differentiate $(q(x))^{n-1}$. (This is the induction hypothesis.) Then, we observe that

$$r(x) = (q(x))^n = q(x)(q(x))^{n-1}.$$

We'll use the product rule to differentiate $r(x)$ where the derivatives of the factors use the base case and the induction hypothesis.

$$r'(x)$$
$$= q'(x)q(x)^{n-1} + q(x)(n-1)(q(x))^{n-2}q'(x)$$
$$= n(q(x))^{n-1}q'(x).$$

Similarly, we can get the full chain rule for polynomials

from Theorem 1.2.3 by using the definition of the composition of polynomials to break any composition into a sum of compositions with individual powers. We'll leave this as an exercise, but we won't actually need more than Theorem 1.2.3 for our purposes.

A remark: This has been a really horrible way of proving the product rule and chain rule. It only works for polynomials. And it makes the whole subject appear as if it is a kind of list of random identities. The moral is that the definition of the derivative and the proof of the rules of differentiation from the definition are useful even if the only functions you will ever differentiate are polynomials, because they make the subject more conceptual. We will cover that in detail later.

Now that we know all about derivatives of polynomials, we can use this information to derive other facts about polynomials which are usually considered more basic. For instance:

Theorem 1.2.4 **The binomial theorem, sort of**   Let x be a variable, y be a rational number, and n be a natural number. Then

$$(x + y)^n = \sum_{j=0}^{n} \binom{n}{j} x^j y^{n-j}.$$

Here we define the *binomial coefficient* $\binom{n}{j}$ by

$$\binom{n}{j} = \frac{n!}{(n-j)!j!}.$$

The hypotheses of our theorem should look a bit comical. Here we're treating x and y as if they aren't symmetrical purely so that we can prove this theorem with the machinery we've already developed. We could fix this of course. (How?) It might involve slightly more tedious definitions. If you see a proof of the binomial theorem in high school, usually it's a discussion of how you expand out the product $(x + y)^n$. The binomial coefficient $\binom{n}{j}$ has a meaning. It counts the number of ways of choosing j positions from n. The j positions we're choosing are the positions in which we have an x. What was just sketched is in some sense the right conceptual proof. Now you'll see a completely different proof.

Proof of the Binomial theorem

First

$$(x+y)^n = \sum_{j=0}^{n} b_{j,n} x^j y^{n-j},$$

where $b_{j,n}$ are some numbers. (Why?) Our job is to show that $b_{j,n} = \binom{n}{j}$. We will do this by differentiating both sides of the equation j times (in the variable x, of course). When we differentiate the right hand side j times, the constant term is

$$j! b_{j,n} y^{n-j}.$$

When we differentiate the left hand side j times using the chain rule, we get

$$\frac{n!}{(n-j)!}(x+y)^{n-j}.$$

The constant term is

$$\frac{n!}{(n-j)!} y^{n-j}.$$

Therefore

$$j! b_{j,n} = \frac{n!}{(n-j)!},$$

or

$$b_{j,n} = \frac{n!}{(n-j)! j!} = \binom{n}{j}.$$

Taylor expansions for functions are going to be a recurring theme throughout the book. Here, we're going to see that all polynomials are equal to their Taylor expansion about each point. Roughly this is the result from high school algebra which says that polynomials can be expanded about any point. I'm not sure how much this is emphasized in high schools. We'll definitely have cause to use it later. For instance, when we develop methods of numerical integration, it will be useful to approximate the function we're integrating by a different polynomial in each of a number of intervals. When we do this, it will make computations simpler to expand the polynomial about the midpoint of each interval.

Theorem 1.2.5 Taylor's theorem for polynomials

Let $p(x) = a_0 + a_1 x + \cdots + a_k x^k$ be a polynomial of degree k. Let y be a rational number. Then

$$p(x) = p(y) + p'(y)(x - y) + \frac{p''(y)}{2}(x - y)^2$$

$$+ \cdots + \frac{p^{(k)}(y)}{k!}(x - y)^k.$$

Proof of Taylor's theorem for polynomials

This is again a job for induction. When the degree $k = 0$, the polynomial is constant and $p(y) = a_0$ for all y. Thus the base case is true. Now we assume the induction hypothesis, that the theorem is true for polynomials of degree $k - 1$. Note that the polynomial

$$q(x) = p(x) - a_k(x - y)^k,$$

is a polynomial of degree $k - 1$ by Theorem 1.2.4. Thus by the induction hypothesis

$$q(x) = p(y) + p'(y)(x - y) + \frac{p''(y)}{2}(x - y)^2$$

$$+ \cdots + \frac{p^{(k-1)}(y)}{(k-1)!}(x - y)^{k-1}.$$

Proof of Taylor's theorem for polynomials, continued

This is because all the derivatives of $(x - y)^k$ up to the $k - 1$st vanish at y so evaluating the derivatives of q at y is the same as evaluating the derivatives of p. Now simply observe that since the kth derivative of p is $k!a_k$, that the last term $a_k(x - y)^k$ is exactly $\frac{p^{(k)}(y)}{k!}(x - y)^k$.

To sum up, we have developed differentiation and its rules in the case of polynomials without referencing in any way the meaning of the derivative. This has allowed us by judicious use of proof by induction to develop a few of the basic elements of the theory of polynomials. We have derived the binomial theorem for polynomials, and we have also derived Taylor's theorem for polynomials. In particular, this allows us to expand polynomials about arbitrary points.

Exercises for Section 1.2

1. Complete the proof of the product rule for differentiation of polynomials given the product rule for differentiation of monomials. (Hint: Expand the product.)

2. Prove the chain rule for differentiation of polynomials: If $r(x) = p(q(x))$ with p and q polynomials, then

$$r'(x) = p'(q(x))q'(x).$$

Hint: Use the case of $r(x) = (q(x))^n$, and expand $p(q(x))$ according to the terms of p.

3. To do this exercise, feel free to use anything you know about derivatives of functions. Prove the following identity for k an odd integer:

$$\frac{d^{k+1}}{dx^{k+1}}[(1+x^2)^{\frac{k}{2}}] = \frac{((1)(3)\ldots(k))^2}{(1+x^2)^{\frac{k+2}{2}}}.$$

Hint: Try induction on k.

4. Prove the generating rule for Pascal's triangle. That is, show for any natural numbers $0 < k < n$ that the formula

$$\binom{n}{k} + \binom{n}{k-1} = \binom{n+1}{k}.$$

Hint: Put both terms of the left hand side under a common denominator. Aside: One interpretation of this result is that when you choose k items from $n + 1$, either you choose the last one or you don't.

5. Use the previous problem to give a proof of the Binomial theorem by induction. Hint: To prove the inductive step, use $(x + y)^{n+1} = (x + y)(x + y)^n$, expand out the first factor and use the induction hypothesis on the second.

6. Setting $j = k - 1$, rewrite the generating rule for Pascal's triangle as

$$\binom{n}{j} = \binom{n+1}{j+1} - \binom{n}{j+1}.$$

Use this equation to prove by induction a formula for

$$\sum_{n=1}^{m} \binom{n}{j}.$$

Hint: It's a telescoping sum.

7. Use the previous exercise to derive a formula for the sum of the first m fourth powers,

$$\sum_{n=1}^{m} n^4.$$

Hint: $\binom{n}{4}$ is a degree 4 polynomial in n. Write n^4 in terms of $\binom{n}{4}, \binom{n}{3}, \binom{n}{2}$, and $\binom{n}{1}$.

Flipped recitation, derivatives of polynomials

In the section on derivatives of polynomials, after developing the rules of differentiation for polynomials, we derived the binomial theorem and Taylor's theorem for polynomials.

Problem 1 The derivative of a product.
Consider

$$f(x) = (x^2 + 3x + 4)(x^3 + 5x + 6).$$

Take the derivative of $f(x)$ using the product rule. Then expand out the product which defines $f(x)$, differentiate, and verify that you got the same answer.

Problem 2 Use the binomial theorem to expand $(x + 7)^6$. Differentiate the expanded expression. Verify that the answer is $6(x + 7)^5$.

Problem 3
Consider the polynomial

$$p(x) = x^5 + 3x^4 + 6x^3 + 4x^2 + 2x + 1.$$

Use Taylor's theorem for polynomials to expand this about $y = 2$. Use the binomial theorem to expand out the expression that you get and verify it is $p(x)$.

◇ 1.3 The real numbers

Now let us develop the real number system. This might seem like a very strange thing for us to be doing. It must seem to you that you have been studying real numbers most of your life. However, some introspection is likely to reveal that not everything you have been told about the real numbers is entirely believable. (For example, a recent seventh grade textbook explains that to add and multiply rational numbers, you should follow a set of rules you have been given. To add and multiply reals, you should plug them into your calculator.)

Because the real numbers will be the central focus of inquiry in this book, we will take this moment to specify exactly what they are. The central tenet of mathematics is that you must always tell the truth, and you can't be sure that you are doing this about real numbers, unless you are sure exactly what they are. There is more than one possible approach to defining them. Most mathematicians focus on what you expect to do with the real numbers. You are given a set of axioms to cover this. It should be possible to perform basic arithmetic on the reals (the field axioms), there should be a way of comparing the size of two real numbers (the total ordering axiom) and a lot of limits should exist (the least upper bound property). After you have written down these axioms, you are in a good position to start proving theorems about the real numbers. To see how specifically this works, you can find one version of the axioms in the Appendix. But you might be quite confused about what is going on. "Are these really the same real numbers you've always heard about?" "Are the real numbers the only set of numbers satisfying these axioms?" " What are individual real numbers like?" It is possible with some work to proceed from the axioms to answer those questions, but the work is nontrivial.

We will take a slightly different approach. We will describe the real numbers in much the way they were described to you in grade school, as decimal expansions. (Mathematicians tend not to like this because there are arbitrary choices like the choice of the base ten.) Then because we'd like to use the real numbers, we will check that they satisfy the axioms allowing us to order them, take limits, and do arithmetic. It will turn out that doing arithmetic is the hardest part. (There's a reason you need a calculator!) You may have been trained to take for granted that arithmetic works on the real numbers just as on any more basic kind of number. But in order to actually be able to perform any arithmetic operation on general real numbers, you have to take limits.

Before we start, perhaps a few words are required about the usefulness of this. You are right to be concerned. As scientists and engineers pursuing practical objectives, you will not encounter any typical real number. Sure, you might collect some data. But it will come to you as floating point numbers with implicit error intervals. Why then should we study something so abstract, so idealized, so *unreal* as the real numbers? The answer is that quite happily, the processes which we use to draw conclusions about real numbers and especially to study their limits (the main subject of the book) are exactly the same as those used to rigorously study floating point numbers with error intervals. It might be wise to take this viewpoint about this entire book. But this requires thinking differently than you are used to about what the main questions are.

Now we begin formally. What is a real number?

The real numbers

A real number is an expression of the form

$$\pm a_1 a_2 a_3 \ldots a_m . b_1 b_2 b_3 \ldots .$$

Here the $\pm$ represents a choice between plus and minus. The leading digit a_1 is an integer between 1 and 9 inclusive. All other digits are integers between 0 and 9 inclusive.

The real numbers cont. One detail is that some real numbers have two such representations, namely a terminating decimal

$$\pm a_1 a_2 \ldots a_m.b_1 b_2 \ldots b_n 000\ldots,$$

where b_n is different from 0, and

$$\pm a_1 a_2 \ldots a_m.b_1 b_2 \ldots (b_n - 1)999\ldots,$$

a decimal with repeating 9's. (Note that the repeating 9's could start to the left of the decimal place just as well as to the right.) The set of real numbers, we will invariably refer to as **R**.

Hopefully, we have now described the real numbers as you have seen them since grade school. It is often pointed out that they can be visualized as populating a line. You can do this by first marking off the integers at equal distances on the line. Then the interval between any two consecutive integers can be cut into ten equal subintervals. The value of the first digit after the decimal describes which interval the real number lies in. You continue the process, subdividing each of those ten intervals into ten equal subintervals and so on.

When dealing with the real numbers in practice, we very often approximate to a few decimal places. Strangely, there is no standard notation for this, so we introduce some.

Truncation Given a real number

$$x = \pm a_1 a_2 \ldots a_m.b_1 b_2 \ldots b_n b_{n+1} b_{n+2} \ldots,$$

we define $t_n(x)$, the truncation to n decimal places, as

$$t_n(x) = \pm a_1 a_2 \ldots a_m.b_1 b_2 \ldots b_n.$$

Truncation cont.

In order for t_n to be a well-defined function on the reals, we must specify how it acts on reals with two decimal representations (the case of repeating 0's and 9's). We specify that to apply t_n, we always take the representation with repeating zeroes. Thus, given any real number, we uniquely map it with t_n to a terminating decimal, which we can also view as a rational number with denominator 10^n. We note that as n increases with x fixed, the truncation $t_n(x)$ increases.

We are now ready to define inequalities among real numbers.

Greater than and less than

Given two real numbers x and y, we say that $x \geq y$ if $x = y$ or there is some n for which $t_n(x) > t_n(y)$. (Ask yourself why we need the inequality between the truncations to be strict.)

When presented with a new definition, it is often valuable to think about it in terms of algorithms. How do we check if the number x is greater than or equal to the number y? If x is actually greater, then we'll find out in a finite number of steps as we find a decimal place n where the truncation of x is actually bigger. If x and y are equal, we'll never find out, because we'd have to check all the truncations. While at first this seems an unhappy state of affairs, it actually agrees with our intuition about approximations and error intervals. If two approximations are far apart so that their error intervals are disjoint, we can tell which one is bigger. Otherwise, we're not sure. Already, we see that in this way, the real numbers, which are an idealization, model reality well.

We now state as a theorem that any two real numbers can be ordered.

Theorem 1.3.1
Given two real numbers x and y, $x \geq y$ or $y \geq x$.

Proof of Theorem 1.3.1
If for some n, we have $t_n(x) > t_n(y)$ or $t_n(x) < t_n(y)$, then we're done. The only case remaining is that $t_n(x) = t_n(y)$ for all n. In this case, x and y have the same decimal expansion and are therefore the same number. If so, both $x \geq y$ and $y \geq x$ hold.

Thus we have completed one third of our project for defining the real numbers. They are ordered. Decimals are in fact quite helpful in the ordering, which is basically alphabetical. (A more technical term for this kind of ordering is *lexicographic*.)

We are now prepared to establish the least upper bound property for the real numbers.

Upper bounds and least upper bounds
Given a set A of real numbers, we say that a real number x is an upper bound for A if for every $y \in A$, we have that $x \geq y$. We say that x is the least upper bound for A if for every other upper bound z for A, we have that $x \leq z$.

This definition will be extremely important to us. If the set A has a maximal element, then the maximal element is the least upper bound. But some sets of real numbers have no maximum. An example is given by $A = (0,1)$, the open unit interval. This set A has no maximum. It contains numbers arbitrarily close to 1, but it doesn't contain 1. But 1 is the least upper bound. The way we see this is that any number x which is strictly less than 1 must satisfy $t_n(x) < t_n(1)$ for some n. Then $t_{n+1}(x) + 10^{-n-1}$ must be in A but greater than x. Thus x is not an upper bound.

We are interested in least upper bounds as a kind of upper limit of the real numbers in the set A. An upper bound might miss being in A by a great deal. A least upper bound will be just outside of A.

**Theorem
1.3.2
Least
upper bound
property**

Any nonempty set of real numbers A which has a real upper bound has a least upper bound in the reals.

**Proof of least
upper bound
property**

We are given that A is nonempty. Let z be an element of it. We are given that it has an upper bound y. Now we are going to find the least upper bound x by constructing its decimal expansion. Since y is an upper bound, so is $a = t_n(y) + \frac{1}{10^n}$. The number a is an upper bound which also has a decimal expansion which terminates at the nth place. Moreover, $a > t_n(z)$. In fact, $10^n(a - t_n(z))$ is a positive natural number. We let B be the set of all natural numbers of the form $10^n(c - t_n(z))$ with c an upper bound for A with decimal expansion terminating at or before the nth place. This set B is a nonempty set of natural numbers which serves as a proxy for the set of upper bounds for A which terminate at n decimal places. To the set B, we may apply the well ordering principle, Theorem 1.1.1 The set B has a smallest element, b. Thus $10^{-n}b + t_n(z)$ is the smallest upper bound for A, with an n-place decimal expansion. We define $x_n = 10^{-n}b + t_n(z) - 10^{-n}$. Thus x_n just misses being an upper bound. If after some finite n, all x_m with $m > n$ are the same, we let x be this x_m. Otherwise, we let x be the real number so that $t_n(x) = x_n$. (We used the well ordering principle to construct the decimal expansion for x. (Why did we treat the case of x with a terminating expansion separately. Hint: It was because of our definition for t_n.)

The above proof may be a little hard to digest. To understand it better, let us consider an example. Often it is touted that one of the virtues of the real number system is that it contains $\sqrt{2}$. (You should be a little concerned that we haven't defined multiplication yet, but this example can be viewed as motivation for the definition.) How do we see that the real numbers contain $\sqrt{2}$? We find a least upper bound for all numbers whose square is less than 2. We do this in the spirit of the above proof. First, we find the smallest number with one decimal place whose square is more than 2. It is 1.5. We subtract 0.1 and record 1.4. Then we find the smallest number with two decimal places whose square is larger than 2. It is 1.42. We subtract 0.01 and record 1.41. Gradually, we build up the decimal expansion for $\sqrt{2}$, which begins 1.414213562. Our algorithm never terminates, but we get an arbitrarily long decimal expansion with a finite number of steps.

The least upper bound property, while it is easy to prove using the decimal system, is a pretty sophisticated piece of mathematics. It is a rudimentary tool for taking limits, something we don't consider in school until we take calculus. Adding and multiplying, though, is one of the first things we think of doing to real numbers. Perhaps we want a calculator to handle it, but we imagine that nothing fancier is going on than our usual algorithms for adding and muliplying. Let's consider how this works. Let's say I want to add two typical real numbers. I write out their decimal expansions one above the other. Then I start at the right. Oops. The numbers have infinite decimal expansions, so I can never get to the right. This problem is not easily waved away. Through the process of carrying, quite insignificant digits of the summands can affect quite significant digits of the sum. In order to calculate, as a practical matter, an arithmetic operation performed on two numbers, we have to take a limit.

Luckily, we have established the least upper bound property. We can use it to define the arithmetic operations on the reals. Our definitions might look circular, but they aren't. Every time we define an operation, we are using the same operation on the rationals in the definition. As long as the numbers we are adding are rational, there's no problem.

Addition and Multiplication

Let x and y be two nonnegative real numbers. We let $A = \{t_n(x) + t_n(y)\}$ be the set of sums of truncations of x and y. We let $M = \{t_n(x)t_n(y)\}$ be the set of products of truncations of x and y. Note that both sets have upper bounds. (We can use $t_n(x) + t_n(y) + \frac{2}{10^n}$ as an upper bound for A [why?] and $(t_n(x) + \frac{1}{10^n})(t_n(y) + \frac{1}{10^n})$ as an upper bound for M. [Why?] Now we apply the least upper bound property to see that A and M have least upper bounds. We define $x + y$ to be the least upper bound for A and xy to be the least upper bound for M. We restricted to x and y positive, so that the expressions $t_n(x) + t_n(y)$ and $t_n(x)t_n(y)$ are increasing in n making the least upper bounds what we really want.

Since we have so far only defined addition and multiplication for positive numbers, defining subtraction of positive numbers seems a high priority.

Subtraction

Again let x and y be nonnegative real numbers. We define $S = \{t_n(x) - t_n(y) - \frac{1}{10^n}\}$. We subtract $\frac{1}{10^n}$ from the nth element so that while we are replacing x by an underestimate $t_n(x)$, we are replacing y by an overestimate $t_n(y) + \frac{1}{10^n}$, and when we subtract, we have an underestimate for the difference. We define $x - y$ to be the least upper bound of S.

What about division?

Division

> Let x and y be nonnegative real numbers. Let $D = \{z : x \geq yz\}$. Thus D consists of real numbers we can multiply by y to get less than x. These are the underestimates of the quotient. We define $\frac{x}{y}$ to be the least upper bound of D.

So how are we doing? The real numbers as we have defined them are the numbers we recognize them from grade school. We have shown that this set of real numbers has an order, that it satisfies the least upper bound property, and that we may perform arithmetic operations. Mathematicians might still not be entirely satisfied as these arithmetic operations still must be proven to satisfy the laws they should inherit from the rational numbers. The proof is not quite as easy as it looks. For instance, let's say we want to prove the distributive law. Thus if x, y, and z are nonnegative real numbers, we would like to show that $\{t_n(x + y)t_n(z)\}$ has the same least upper bound as $\{t_n(xz) + t_n(yz)\}$. It is true and it can be done. But to do it, it helps to deal carefully with something we have completely set aside thus far. It helps to have estimates on how far away a truncated version of $(x + y)z$ actually is from the least upper bound.

This gets at an objection that a practical person could have for the way we've defined our operations thus far. Certainly, the least upper bounds exist. But they are the output of an algorithm that never terminates. To actually use real numbers as a stand in for approximations with error intervals, we need to be able at each step of a never terminating algorithm to have control on the error. Notice we did have that kind of control in the example with $\sqrt{2}$. When we had n decimal places, we knew we were within 10^{-n} of the answer. In the next section, we will get at both the practical and theoretical versions of this problem by introducing the definition of the limit. We will see that understanding that a limit exists is more than knowing what the limit is. It also involves estimating how fast the limit converges. In practical terms, this means calculating an error interval around the limitand.

Let us sum up what we have accomplished so far. We defined the set of real numbers precisely in terms of decimal expansions. We defined the order of the real numbers in terms of truncations of the decimal expansions. We showed that the reals satisfy the least upper bound property, which gives us a kind of rudimentary limiting process. Using that limiting process, we are able to define the arithmetic operations on the real numbers. You should view this as a somewhat impressive accomplishment. You have been studying real numbers and their operations for some time, but it may be that no one has ever defined them for you. Only now, with the definitions in hand, do you have the power to check whether things you have been told about the real numbers are true.

Exercises for Section 1.3

1. We defined the difference $x - y$ of two real numbers x and y in terms of differences between the decimal truncations $t_n(x)$ and $t_n(y)$ of the real numbers. We could instead have done the following. Define the set $S_{x,y}$ to be the set of rational numbers $p - q$ so that $p < x$ and $q > y$. Show that for any real numbers x and y, the set $S_{x,y}$ is bounded above. Show that the least upper bound of $S_{x,y}$ is $x - y$.

2. The purpose of this exercise is to show that when considering the sum $x + y$ of two real numbers x and y, if we are only interested in knowing $t_n(x + y)$, the truncation of the sum up to n places past the decimal, we may need information about x and y arbitrarily far into their decimal expansion. Let us make this precise. Fix a natural number n. Now choose another, possibly much larger natural number m. Show that you can find real numbers x_1, x_2, y_1, and y_2 with the properties that

$$t_m(x_1) = t_m(x_2),$$

$$t_m(y_1) = t_m(y_2),$$

but

$$t_n(x_1 + y_1) \neq t_n(x_2 + y_2).$$

3. The purpose of this exercise is to show that when considering the product xy of two real numbers x and y, if we are only interested in knowing $t_n(xy)$, the truncation of the sum up to n places past the decimal, we may need information about x and y arbitrarily far into their decimal expansion. Let us make this precise. Fix a natural number n. Now choose another, possibly much larger natural number m. Show that you can find positive real numbers x_1, x_2, y_1, and y_2 with the properties that

$$t_m(x_1) = t_m(x_2),$$

$$t_m(y_1) = t_m(y_2),$$

but

$$t_n(x_1 y_1) \neq t_n(x_2 y_2).$$

Hint: If the product is really close to a fraction with denominator 10^n, very small changes in the factors can get you on either side of it.

4. Let x be a positive real number with a repeating decimal expansion. This means that if

$$x = a_m a_{m-1} \ldots a_1.b_1 b_2 \ldots b_m \ldots,$$

there is some natural number N and some natural number j so that when $k \geq N$, we always have $b_{k+j} = b_k$. We call j the period of repetition. So, for example, if $x = \frac{1}{7} = 0.142857142857\ldots$ then we have $N = 1$ and $j = 6$. Show that any x with a repeating decimal expansion is a rational number. That is, show that it is the quotient of two integers. (Hint: Compare $10^j x$ with x. [But remember that to do this exercise correctly, you have to use the definition of division.])

5. Let z and w be rational numbers having denominator 10^n (not necessarily in lowest terms). Consider all possible real x with $t_n(x) = z$ and all possible real y with $t_n(y) = w$. What are all possible values of $t_n(x+y)$ in terms of z and w?

6. Let x, y, and z be real numbers. Suppose $x \leq y$ and $y \leq z$. Show that $x \leq z$. Hint: Use the definition of $\leq$, of course. You are allowed to use that the statement is true when x, y, and z are rational. Your proof may break into cases because there are two ways for the definition to be satisfied.

7. The purpose of this exercise is to check a detail in the proof of the least upper bound property. Let A be a nonempty set of real numbers which is bounded above. Suppose that A does not have a terminating decimal expansion as its least upper bound. Let x_n be the largest decimal which terminates at the nth place and is not an upper bound for A. (This agrees with the notation in the proof of the Least Upper Bound property above.) Show that when n and m are natural numbers with $n < m$, then $t_n(x_m) = x_n$. Conclude that there is a real number x with $t_n(x) = x_n$. Show that x is an upper bound for A. Conclude that x is the least upper bound. Hint: The first part is true without the hypothesis that A does not have a terminating least upper bound. You have to use this hypothesis for the second part, however. You can specify the decimal expansion of x, but for it to have the desired property, it must be that the decimal expansion you specify does not have repeating 9's. To prove that x is the least upper bound, use proof by contradiction. Suppose that y is a strictly smaller upper bound. Then there exists n with $t_n(y) < t_n(x)$. Reach a contradiction.

Flipped recitation, real numbers

In the section on real numbers, we precisely defined the real numbers and their order, and we proved the least upper bound property and used it to define operations

Problem 1 Is multiplication really defined?
Let x and y be positive real numbers. Let

$$A = \{t_n(x)t_n(y) : n \in \mathbf{N}\}.$$

Show that for any positive integer m, we have that $(t_m(x) + \frac{1}{10^m})(t_m(y) + \frac{1}{10^m})$ is an upper bound for A.

Problem 2 Greatest lower bound.

Let A be a nonempty set of reals which is bounded below. (That is, there is a real number y so that $x \geq y$ for every $x \in A$. Such a y is called a lower bound.) Show that A has a greatest lower bound. Hint: Flip everything upside down.

Problem 3

Let

$$A = \{(1 + \frac{1}{n})^n : n \in \mathbf{N}\}.$$

Show that A has an upper bound. Hint: Use the binomial theorem. Don't be afraid to lose some ground.

◇ 1.4   Limits

In the previous section, we used the least upper bound property of the real numbers to define the basic arithmetic operations of addition and multiplication. In effect, this involved finding sequences which converged to the sum and product. In the current section, we will make the notion of convergence to a limit by a sequence more flexible and more precise. In general, a sequence of real numbers is a set of real numbers $\{a_n\}$ which is indexed by the natural numbers. That is each element of the sequence a_n is associated to a particular natural number n. We will refer to a_n as the nth element of the sequence.

Example 1   A sequence converging to a product

Let x and y be positive real numbers. Let $t_n(x)$ and $t_n(y)$ be as defined in section 1.3, the truncations of x and y to their decimal expansions up to n places. Consider the sequence $\{a_n\}$ given by

$$a_n = t_n(x)t_n(y).$$

The number a_n represents the approximation to the product xy obtained by neglecting all contributions coming from parts of x or parts of y after the nth decimal place. The sequence a_n is increasing, meaning that if $n > m$ then $a_n \geq a_m$.

Example 2   A sequence converging to 1 whose least upper bound is not 1

Consider the sequence $\{b_n\}$ given by

$$b_n = 1 + (-2)^{-n}.$$

The sequence $\{b_n\}$ is neither increasing nor decreasing since the odd elements of the sequence are less than 1 while the even ones are greater than 1.

As a proxy for taking a limit of the sequence in Example 1, when we studied it in Section 1.3, we took the least upper bound. But this only worked because the sequence was increasing. In Example 2, the least upper bound is $\frac{5}{4}$ since we have written 0 out of the natural numbers. The greatest lower bound is $\frac{1}{2}$. But the limit should be 1 since the sequence oscillates ever closer to 1 as n increases. In what follows, we will write down a definition of convergence to limits under which both sequences converge. (This should not be too surprising.) Indeed, if you have already studied calculus, you probably have a good sense about which sequences converge. Nevertheless, you should pay careful attention to the definition of convergence, because though it is technical, it contains within it a way of quantifying not just whether sequences converge but how fast. This is information of great practical importance.

Sequence converging to a limit   We say that the sequence $\{a_n\}$ converges to the limit L if for every real number $\epsilon > 0$ there is a natural number N so that $|a_n - L| < \epsilon$ whenever $n > N$. We sometimes denote L by

$$\lim_{n \to \infty} a_n.$$

Example 3   The limit of the sequence in Example 2

Let $\{b_n\}$ be as before. We will show that the sequence $\{b_n\}$ converges to the limit 1. We observe that $|b_n - 1| = 2^{-n}$. To complete our proof, we must find for each real number $\epsilon > 0$ a natural number N so that when $n > N$, we have that $2^{-n} < \epsilon$. Since the error 2^{-n} decreases as n increases, it is enough to ensure that $2^{-N} < \epsilon$. We can do this by taking $N > \log_2 \frac{1}{\epsilon}$. A note for sticklers: If we want to be completely rigorous, maybe we have to verify that $\log_2$ is defined for all real numbers, which we haven't done yet. However, it is quite easy to see that for any $\epsilon > 0$, there is an m so that $10^{-m} < \epsilon$. This is readily done since ϵ being nonzero and positive has a nonzero digit in its decimal expansion and we can choose m to be the place of that digit. Then we can use an inequality like $2^{-4m} = 16^{-m} < 10^{-m}$.

It is possible to study proofs and fall under the impression that proofs are a bunch of verbiage used to certify some trivial fact that we already know. They have to be written in complete and grammatical sentences and follow basic rules of logic. All of this can be said to be true about proofs that limits exist. But there is one additional element that you have to supply in order to prove a limit exists. You have to find a function $N(\epsilon)$, since the choice of the number N depends on the number ϵ. What does this function mean? Here ϵ is a small number. It is the error we are willing to tolerate between a term in the sequence and the limit of the sequence. Then $N(\epsilon)$ represents how far we need to go in the sequence until we are certain that the terms in the sequence will approximate the limit to our tolerance. This can be very much a practical question. For instance in the first example, the terms in the sequence are approximations to the product xy which can be calculated in a finite number of steps. Recall that there exist seventh grade textbooks which insist that real numbers be multiplied by calculators which seems ridiculous since calculators can only show numbers up to some accuracy ϵ (which used to be 10^{-8}.) In order for the calculator to comply with the requirements of the seventh grade textbook it needs to know $N(10^{-8})$ so that it will know how far in the sequence it has to go to get an answer with appropriate accuracy.

Thus the function $N(\epsilon)$ is really important. It is strange that it is so easy to think that proofs that limits exist only answer questions to which you already know the answer. This is because you might have superb intuition as to whether limits exist. But why does the question always have to be "Does the limit exist?" Couldn't it equally well be "What is a function $N(\epsilon)$ which certifies that the limit exists?" The standard choice of question is because of deep anti-mathematical biases in society. After all, the first question has a unique yes or no answer. For the second question, the answer is a function and it is not unique. In fact, given a function that works, any larger function also works. Of course, it can also be said that answers to the second question, even if correct, do not have equal value. It is better for the calculator to get as small a function $N(\epsilon)$ as it can that it can guarantee works.

You will be required to prove limits exist as you proceed. For some tips about how to write grammatically and logically, see the appendix. I won't discuss it further in the body of the text. But a really legitimate question for you to be asking yourselves is "How do I find a function $N(\epsilon)$?" The most obvious thing to say is that you should be able to estimate the error between the Nth term of the sequence and the limit. If this error is decreasing, you have already found the inverse function for $N(\epsilon)$ and just have to invert. (This is what happened in the third example: the function $\log_2(\frac{1}{\epsilon})$ is the inverse of the function 2^{-n}.) If the errors aren't always decreasing, you may have to get an upper bound on all later errors too.

The question still remains, how do we find these upper bounds? Therein lies the artistry of the subject. Because we are estimating the difference between a limit and a nearby element of a sequence, there is often a whiff of differential calculus about the process. This may seem ironic since we have not yet established any of the theorems of differential calculus and this is one of our goals. Nevertheless, your skills at finding derivatives, properly applied, may prove quite useful.

Example 4   The limit of the sequence in Example 1

We would like to show that for x and y positive real numbers, the sequence $\{t_n(x)t_n(y)\}$ converges to the product xy, which is defined as the least upper bound of the sequence. Thus we need to estimate $|xy - t_n(x)t_n(y)| = xy - t_n(x)t_n(y)$. We observe that

$$t_n(x)t_n(y) \le xy \le (t_n(x) + \frac{1}{10^n})(t_n(y) + \frac{1}{10^n}),$$

since $t_n(x) + \frac{1}{10^n}$ has a larger nth place than any truncation of x and similarly for y. Now subtracting $t_n(x)t_n(y)$ from the inequality, we get

$$0 \le xy - t_n(x)t_n(y) \le (t_n(x) + \frac{1}{10^n})(t_n(y) + \frac{1}{10^n}) - t_n(x)t_n(y).$$

Note that the right-hand side looks a lot like the expressions we get from the definition of the derivative, where $\frac{1}{10^n}$ plays the role of h. Not surprisingly then, when we simplify, what we get is reminiscent of the product rule,

$$0 \le xy - t_n(x)t_n(y) \le \frac{1}{10^n}(t_n(x) + t_n(y) + \frac{1}{10^n}) \le \frac{1}{10^n}(x + y + 1).$$

Note that we are free to use the distributive law because we are only applying it to rational numbers. The last step is a little wasteful, but we have done it to have a function that is readily invertible. Clearly $\frac{1}{10^n}(x + y + 1)$ is decreasing as n increases. Thus if we just solve for N in

$$\epsilon = \frac{1}{10^N}(x + y + 1),$$

we find the function $N(\epsilon)$. It is easy to see that $N(\epsilon) = \log_{10} \frac{x+y+1}{\epsilon}$ works. To summarize the logic, when $n > N(\epsilon)$, then

$$|xy - t_n(x)t_n(y)| \le \frac{1}{10^n}(x + y + 1) \le \epsilon.$$

Thus we have shown that $t_n(x)t_n(y)$ converges to the limit xy.

A clever reader might think that the hard work of the example above is really unnecessary. Shouldn't we know just from the fact that the sequence is increasing that it must converge to its least upper bound? This is in fact the case.

Theorem 1.4.1
Least upper bounds are limits   Let $\{a_n\}$ be an increasing sequence of real numbers which is bounded above. Let L be the least upper bound of the sequence. Then the sequence converges to the limit L.

Proof that Least upper bounds are limits

We will prove the theorem by contradiction. We suppose that the sequence $\{a_n\}$ does not converge to L. This means there is some real number $\epsilon > 0$ for which there is no N, so that when $n > N$, we are guaranteed that $L - \epsilon \le a_n \le L$. This means there are arbitrarily large n so that $a_n < L - \epsilon$. But since a_n is an increasing sequence, this means that all $a_n < L - \epsilon$, since we can always find a later term in the sequence, larger than a_n which is smaller than $L - \epsilon$. We have reached a contradiction since this means that $L - \epsilon$ is an upper bound for the sequence, so L cannot be the least upper bound.

A direct application of the theorem shows that the limit of the first example converges. Is the clever reader right that the fourth example is unnecessary? Not necessarily. A practical reader should object that the proof through the theorem is entirely unquantitative. It doesn't give us an explicit expression for $N(\epsilon)$ and so it doesn't help the calculator one iota. It provides no guarantee of when the approximation is close to the limit. Mathematicians are known for looking for elegant proofs, where "elegant" is usually taken to mean short. In this sense, the proof through Theorem 1.4.1 is elegant. That doesn't necessarily make it better. Sometimes if you're concerned about more than what you're proving, it might be worthwhile to have a longer proof, because it might give you more information.

Example 5 Do the reals satisfy the distributive law?

Let x,y,z be positive real numbers. We would like to show that $(x+y)z = xz+yz$. Precisely, this means that we want to show that

$$\lim_{n\to\infty} t_n(x+y)t_n(z) = \lim_{n\to\infty} t_n(x)t_n(z) + \lim_{n\to\infty} t_n(y)t_n(z).$$

If $L_1 = (x+y)z$, $L_2 = xz$, and $L_3 = yz$, then these are the limits in the equation above. From the definition of the limit, we can find an N_1 so that for $n > N_1$ the following three inequalities hold:

$$|L_1 - t_n(x + y)t_n(z)| \leq \frac{\epsilon}{4}.$$

$$|L_2 - t_n(x)t_n(z)| \leq \frac{\epsilon}{4},$$

and

$$|L_3 - t_n(y)t_n(z)| \leq \frac{\epsilon}{4}.$$

Basically, we find an N for each inequality and take N_1 to be the largest of the three. To get N_1 explicitly, we can follow the previous example.

Next we observe that

$$(t_n(x)+t_n(y))t_n(z) \leq t_n(x+y)t_n(z) \leq (t_n(x)+t_n(y)+\frac{2}{10^n})t_n(z),$$

since the right-hand side is more than $x + y$. Thus

$$(t_n(x)+t_n(y))t_n(z) \leq t_n(x+y)t_n(z) \leq (t_n(x)+t_n(y)+\frac{2}{10^n})t_n(z),$$

from which we can conclude (again following the ideas of the previous example) that there is N_2 so that when $n > N_2$ we have

$$|t_n(x + y)t_n(z) - (t_n(x) + t_n(y))t_n(z)| \leq \frac{\epsilon}{4}.$$

Now take N to be the maximum of N_1 and N_2. We have shown that when we go far enough in each sequence past N, the terms in limiting sequences to L_1, L_2, and L_3 are within $\frac{\epsilon}{4}$ of the limits and that the difference between the nth term in the sequence for L_1 and the sum of the nth terms for L_2 and L_3 is at most $\frac{\epsilon}{4}$. Combining all the errors, we conclude that

$$|L_1 - L_2 - L_3| \leq \epsilon.$$

Since ϵ is an arbitrary positive real number and absolute values are nonnegative, we conclude that $L_1 - L_2 - L_3 = 0$, which is what we were to show.

It is worth noting that when we combined the errors, we were in effect applying the triangle inequality,

$$|a - c| \leq |a - b| + |b - c|$$

multiple times. This inequality holds for all reals a,b,c.

In the spirit of the last example, we are in a position to establish for the reals all arithmetic identities that we have for the rationals. Basically we approximate any quantity we care about closely enough by terminating decimal expansions and we can apply the identity for the rationals. For this reason, we will not have much further need to refer to decimal expansions in the book. We have established what the real numbers are and that they do what we expect of them. Moreover, we have seen how to use the formal definition of the limit and what it means. In what follows, we will discuss additional criteria under which we are guaranteed that a limit exists.

Exercises for Section 1.4

1. Let a, b, c be real numbers. Show that

$$|a - c| \leq |a - b| + |b - c|.$$

   Hint: For rationals, it is enough just to compare the signs. But you can view reals as limits of rationals

2. Use the definition of multiplication of real numbers to show that multiplication of real numbers is associative. That is, show that for any real numbers x, y, z, one has the identity

$$x(yz) = (xy)z.$$

3. Let x and y be real numbers. Suppose that for every $\epsilon > 0$, you have $x < y + \epsilon$. Show that $x \leq y$. Hint: Compare the truncations of x and y.

4. Define the square root function on the positive real numbers by letting $\sqrt{x}$ be the least upper bound of $\{y : y^2 < x\}$, the set of reals whose square is less than x. Prove using the definition of multiplication that the product of $\sqrt{x}$ with itself is x. Hint: It is easy to see from definitions (but you have to show it) that $(\sqrt{x})^2 \leq x$. Use the previous exercise to show also that $(\sqrt{x})^2 \geq x$.

5. Prove using the definition of the limit of a sequence that

$$\lim_{n \to \infty} \sqrt{1 + \frac{1}{n}} = 1.$$

   Give an explicit expression for the function $N(\epsilon)$ that you use. Hint: Compare $\sqrt{1 + \frac{1}{n}}$ with $1 + \frac{1}{2n}$.

6. For the purposes of this exercise, when x is a positive real number define $(x)^{\frac{1}{4}}$ to be the least upper bound of $\{y > 0 : y^4 < x\}$. Prove using the definition of the limit of a sequence that

$$\lim_{n \to \infty} \left(1 + \frac{1}{n}\right)^{\frac{1}{4}} = 1.$$

   Give an explicit expression for the function $N(\epsilon)$ that you use. Hint: Compare $(1 + \frac{1}{n})^{\frac{1}{4}}$ with $1 + \frac{1}{4n}$. Use the definition of the fourth root to make this comparison.

Flipped recitation, limits
We defined the limit of a sequence. We used this definition to establish the distributive law for real numbers.

Problem 1
Calculate

$$\lim_{n \to \infty} (3 - \frac{1}{n})^2,$$

and prove the sequence converges to the limit.

Problem 2
Calculate

$$\lim_{n \to \infty} \frac{1}{5 - \frac{1}{n}}$$

and prove the sequence converges to the limit.

Problem 3
Calculate

$$\lim_{n \to \infty} n(\sqrt{2 - \frac{1}{n}} - \sqrt{2}),$$

and prove the sequence converges to the limit.

Chapter 2

SEQUENCES AND SERIES

The purpose of this section is modest. It is to state certain conditions under which we are guaranteed that limits of sequences converge.

Cauchy sequence

We say that a sequence of real numbers $\{a_n\}$ is a *Cauchy sequence* provided that for every $\epsilon > 0$, there is a natural number N so that when $n, m \geq N$, we have that $|a_n - a_m| \leq \epsilon$.

Example 1 **Given a real number x, its sequence of truncations $\{t_n(x)\}$ is a Cauchy sequence.**

Proof

If $n, m \geq N$, we have that $|t_n(x) - t_m(x)| \leq 10^{-N}$, since they share at least the first N places of their decimal expansion. Given any real number $\epsilon > 0$, there is an $N(\epsilon)$ so that $10^{-N(\epsilon)} < \epsilon$. Thus the sequence $\{t_n(x)\}$ is a Cauchy sequence.

The above example was central in our construction of the real numbers. We got the least upper bound property by associating to each sequence of truncations the real number x which is its limit. The class of Cauchy sequences should be viewed as a minor generalization of the example, as the proof of the following theorem will indicate.

Theorem 2.1.1   Every Cauchy sequence of real numbers converges to a limit.

Proof of Theorem 2.1.1   Let $\{a_n\}$ be a Cauchy sequence. For any j, there is a natural number N_j so that whenever $n,m \geq N_j$, we have that $|a_n - a_m| \leq 2^{-j}$. We now consider the sequence $\{b_j\}$ given by

$$b_j = a_{N_j} - 2^{-j}.$$

Notice that for every n larger than N_j, we have that $a_n > b_j$. Thus each b_j serves as a lower bound for elements of the Cauchy sequence $\{a_n\}$ occurring later than N_j. Each element of the sequence $\{b_j\}$ is bounded above by $b_1 + 1$ for the same reason. Thus the sequence $\{b_j\}$ has a least upper bound, which we denote by L. We will show that L is the limit of the sequence $\{a_n\}$. Suppose that $n > N_j$. Then

$$|a_n - L| < 2^{-j} + |a_n - b_j| = 2^{-j} + a_n - b_j \leq 3(2^{-j}).$$

For every $\epsilon > 0$ there is $j(\epsilon)$ so that $3(2^{-j}) < \epsilon$, and we simply take $N(\epsilon)$ to be $N_{j(\epsilon)}$.

The idea of the proof of Theorem 2.1.1 is that we recover the limit of the Cauchy sequence by taking a related least upper bound. So we can think of the process of finding the limit of the Cauchy sequence as specifying the decimal expansion of the limit, one digit at a time, as this is how the proof of the least upper bound property works.

The converse of Theorem 2.1.1 is also true.

Theorem 2.1.2

Let $\{a_n\}$ be a sequence of real numbers converging to a limit L. Then the sequence $\{a_n\}$ is a Cauchy sequence.

Proof of Theorem 2.1.2

Since $\{a_n\}$ converges to L, for every $\epsilon > 0$, there is an $N > 0$ so that when $j > N$, we have

$$|a_j - L| \leq \frac{\epsilon}{2}.$$

(The reason we can get $\frac{\epsilon}{2}$ on the right-hand side is that we put $\frac{\epsilon}{2}$ in the role of ϵ in the definition of the limit.) Now if j and k are both more than N, we have $|a_j - L| \leq \frac{\epsilon}{2}$ and $|a_k - L| \leq \frac{\epsilon}{2}$. Combining these using the triangle inequality, we get

$$|a_j - a_k| \leq \epsilon,$$

so that the sequence $\{a_j\}$ is a Cauchy sequence as desired.

Combining Theorems 2.1.1 and 2.1.2, we see that what we have learned is that Cauchy sequences of real numbers and convergent sequences of real numbers are the same thing. But the advantage of the Cauchy criterion is that to check whether a sequence is Cauchy, we don't need to know the limit in advance.

Example 2   The following series (that is, infinite sum) converges.

$$S = \sum_{n=1}^{\infty} \frac{1}{n^2}.$$

Proof We may view this series as the limit of the sequence of partial sums

$$a_j = \sum_{n=1}^{j} \frac{1}{n^2}.$$

We can show that the limit converges using Theorem 2.1.1 by showing that $\{a_j\}$ is a Cauchy sequence. Observe that if $j,k >$

N, we definitely have

$$|a_j - a_k| \leq \sum_{n=N}^{\infty} \frac{1}{n^2}.$$

It may be difficult to get an exact expression for the sum on the right, but it is easy to get an upper bound.

$$\sum_{n=N}^{\infty} \frac{1}{n^2} \leq \sum_{n=N}^{\infty} \frac{1}{n(n-1)} = \sum_{n=N}^{\infty} \frac{1}{n-1} - \frac{1}{n}.$$

The reason we used the slightly wasteful inequality, replacing $\frac{1}{n^2}$ by $\frac{1}{n^2-n}$ is that now the sum on the right telescopes, and we know it is exactly equal to $\frac{1}{N-1}$. To sum up, we have shown that when $j,k > N$, we have

$$|a_j - a_k| \leq \frac{1}{N-1}.$$

Since we can make the right-hand side arbitrarily small by choosing N sufficiently large, we see that $\{a_j\}$ is a Cauchy sequence. This example gives an indication of the power of the Cauchy criterion. You would not have found it easier to prove that the limit exists had you known in advance that the series converges to $\frac{\pi^2}{6}$.

Let $\{a_n\}$ be a sequence of real numbers. Let $\{n_k\}$ be a strictly increasing sequence of natural numbers. We say that $\{a_{n_k}\}$ is a subsequence of $\{a_n\}$. We will now prove an important result which helps us discover convergent sequences in the wild.

Theorem 2.1.3 (Bolzano-Weierstrass)   Let $\{a_n\}$ be a bounded sequence of real numbers. (That is, suppose there is a positive real number B so that $|a_j| \leq B$ for all j.) Then $\{a_n\}$ has a convergent subsequence.

Proof of Bolzano-Weierstrass theorem

All the terms of the sequence live in the interval

$$I_0 = [-B,B].$$

We cut I_0 into two equal halves (which are $[-B,0]$ and $[0,B]$). At least one of these contains an infinite number of terms of the sequence. We choose a half which contains infinitely many terms and we call it I_1. Next, we cut I_1 into two halves and choose a half containing infinitely many terms, calling it I_2. We keep going. (At the jth step, we have I_j containing infinitely many terms and we find a half, I_{j+1} which also contains infinitely many terms.) We define the subsequence $\{a_{j_k}\}$ by letting a_{j_k} be the first term of the sequence which follows $a_{j_1}, \ldots, a_{j_{k-1}}$ and which is an element of I_k. We claim that $\{a_{j_k}\}$ is a Cauchy sequence. Let's pick $k,l > N$. Then both a_{j_k} and a_{j_l} lie in the interval I_N, which has length $\frac{B}{2^{N-1}}$. Thus

$$|a_{j_k} - a_{j_l}| \leq \frac{B}{2^{N-1}}.$$

We can make the right-hand side arbitrarily small by making N sufficiently large. Thus we have shown that the subsequence is a Cauchy sequence and hence convergent.

A question you might ask yourself is: How is the proof of the Bolzano-Weierstrass theorem related to decimal expansions?

The final topic in this section is the Squeeze theorem. It is a result that allows us to show that limits converge by comparing them to limits that we already know converge.

Theorem 2.1.4 Squeeze theorem

Given three sequences of real numbers, $\{a_n\}$, $\{b_n\}$, and $\{c_n\}$, if we know that $\{a_n\}$ and $\{b_n\}$ both converge to the same limit L and we know that for each n we have

$$a_n \leq c_n \leq b_n,$$

then the sequence $\{c_n\}$ also converges to the limit L.

Proof of Squeeze theorem

Fix $\epsilon > 0$. There is $N_1 > 0$ so that when $n > N_1$, we have

$$|a_n - L| \leq \epsilon.$$

There is $N_2 > 0$ so that when $n > N_2$, we have

$$|b_n - L| \leq \epsilon.$$

We pick N to be the larger of N_1 and N_2. For $n > N$, the two inequalities above, we know that $a_n, b_n \in (L - \epsilon, L + \epsilon)$. But by the inequality

$$a_n \leq c_n \leq b_n,$$

we know that $c_n \in [a_n, b_n]$. Combining the two facts, we see that

$$c_n \in (L - \epsilon, L + \epsilon),$$

so that

$$|c_n - L| \leq \epsilon.$$

Thus the sequence $\{c_n\}$ converges to L as desired.

Example 3

Calculate

$$\lim_{n \longrightarrow \infty} (1 + \frac{n}{n+1})^{\frac{1}{n}}.$$

The limit above seems a little complicated, so we invoke the squeeze theorem. We observe that the term inside the parentheses is between 1 and 2. (Actually it is getting very close to 2 as n gets large.) Thus

$$1^{\frac{1}{n}} \leq (1 + \frac{n}{n+1})^{\frac{1}{n}} \leq 2^{\frac{1}{n}}.$$

Thus we will know that

$$\lim_{n \to \infty} (1 + \frac{n}{n+1})^{\frac{1}{n}} = 1,$$

provided we can figure out that

$$\lim_{n \to \infty} 1^{\frac{1}{n}} = 1,$$

and

$$\lim_{n \to \infty} 2^{\frac{1}{n}} = 1.$$

The first limit is easy since every term of the sequence is 1. It seems to us that the nth roots of 2 are getting closer to 1, but how do we prove it? Again, it seems like a job for the squeeze theorem. Observe that

$$(1 + \frac{1}{n})^n \geq 2,$$

since $1 + 1$ are the first two terms in the binomial expansion. Thus

$$2^{\frac{1}{n}} \leq 1 + \frac{1}{n}.$$

We know that

$$\lim_{n \to \infty} 1^{\frac{1}{n}} = 1,$$

and perhaps we also know that

$$\lim_{n \to \infty} 1 + \frac{1}{n} = 1,$$

since $\frac{1}{n}$ becomes arbitrarily small as n gets large. Thus by the squeeze theorem, we know

$$\lim_{n \to \infty} 2^{\frac{1}{n}} = 1,$$

and hence

$$\lim_{n \to \infty} (1 + \frac{n}{n+1})^{\frac{1}{n}} = 1.$$

The above example is a reasonable illustration of how the squeeze theorem is always used. We might begin with a very complicated limit, but as long as we know the size of the terms concerned, we can compare, using inequalities, to a much simpler limit.

As of yet, we have not said anything about infinite limits.

Infinite limit   We say that a sequence $\{a_n\}$ of positive real numbers converges to infinity if for every $M > 0$, there is an N so that when $n > N$, we have $a_n > M$. Here M takes the role of ϵ. It is measuring how close the sequence gets to infinity. There is a version of the squeeze theorem we can use to show limits go to infinity.

Theorem 2.1.5 Infinite squeeze theorem   Let $\{a_n\}$ be a sequence of positive real numbers going to infinity. Suppose for every n, we have

$$b_n \geq a_n.$$

Then the sequence $\{b_n\}$ converges to infinity.

Proof of the Infinite squeeze theorem   For every M, there exists N so that when $n > N$, we have $a_n > M$. But since $b_n \geq a_n$, it is also true that $b_n > M$. Thus $\{b_n\}$ goes to infinity.

Example 4   **Show that**

$$\sum_{n=1}^{\infty} \frac{1}{n} = \infty.$$

Proof We will prove this by comparing each reciprocal to the largest power of 2 smaller than it. Thus

$$1+\frac{1}{2}+\frac{1}{3}+\frac{1}{4}+\frac{1}{5}+\frac{1}{6}+\frac{1}{7}+\frac{1}{8}+\cdots > 1+\frac{1}{2}+\frac{1}{4}+\frac{1}{4}+\frac{1}{8}+\frac{1}{8}+\frac{1}{8}+\frac{1}{8}+\cdots.$$

Combining like terms, we get

$$1 + \frac{1}{2} + \frac{1}{3} + \frac{1}{4} + \frac{1}{5} + \frac{1}{6} + \frac{1}{7} + \frac{1}{8} + \cdots > 1 + \frac{1}{2} + \frac{1}{2} + \frac{1}{2} + \cdots.$$

On the right-hand side, we are summing an infinite number of $\frac{1}{2}$'s. Thus the sum is infinite.

Something to think about: The standard proof that the harmonic series (that is, the sum in the previous example) diverges goes by comparing it to the integral of $\frac{1}{x}$ which is a logarithm. Are there any logarithms hiding in the above example?

Exercises for Section 2.1

1. Let $\{a_n\}$ and $\{b_n\}$ be two Cauchy sequences of real numbers. Suppose that for every j, the inequality $|a_j - b_j| \leq \frac{1}{j}$ holds. Show using the definition of the limit of a sequence that the two sequences converge to the same limit.

2. Let C be a subset of the real numbers consisting of those real numbers x having the property that every digit in the decimal expansion of x is 1, 3, 5, or 7. Let $\{c_n\}$ be a sequence of elements of C so that $|c_j| < 1$ for every natural number j. Show that there is a subsequence of $\{c_n\}$ which converges to an element of C.

3. Let x be a positive real number. Show that $\{\sqrt{t_n(x)}\}$ is a Cauchy sequence. Show that the limit is $\sqrt{x}$.

4. Use the squeeze theorem to calculate

$$\lim_{n \to \infty} (1 + \frac{1}{n^2})^n.$$

Hint: For the upper bound, expand using the binomial theorem. Then use the inequality $\binom{n}{j} \leq n^j$. Finally use the identity

$$1 + \frac{1}{n} + \frac{1}{n^2} + \cdots + \frac{1}{n^n} = \frac{1 - \frac{1}{n^{n+1}}}{1 - \frac{1}{n}}.$$

5. Use the squeeze theorem to calculate

$$\lim_{n \to \infty} n(\sqrt{4 + \frac{3}{n}} - 2).$$

Hint: Approximate the square root as a linear expression L in $\frac{1}{n}$ so that the first two terms of the binomial expansion for L^2 are exactly $4 + \frac{3}{n}$. Use L as an upper bound and then correct L by subtracting a multiple of the square of $\frac{1}{n}$ to get a lower bound.

Flipped recitation: Cauchy sequences, Bolzano-Weierstrass, and Squeeze theorems

In this section, we developed a number of conditions under which we would be guaranteed that sequences converge.

Problem 1

Let y be a fixed real number and define the function $f(x) = \frac{x+y}{2}$. Fix a real number a_1 and define the sequence $a_j = f(a_{j-1})$, for $j \geq 2$. Show that $\{a_j\}$ is a Cauchy sequence. Hint: Observe that for any real numbers z, w you have $|f(z) - f(w)| \leq \frac{|z-w|}{2}$.

Problem 2

Let $\{a_j\}$ be a bounded sequence of real numbers. Fix a positive natural number n. Show that there is some value of $t_n(a_j)$ which arises for infinitely many choices of j. How is this related to the Bolzano-Weierstrass theorem?

Problem 3

Show that

$$\sum_{j=1}^{\infty} \frac{1}{(j+1)\log(j+1)}$$

has an infinite limit.

◇ 2.2 Infinite series

In this section, we will restrict our attention to infinite series, which we will view as special kinds of sequences. We will bring what we learned about the convergence of sequences to bear on infinite series.

Infinite series An infinite series is a formal sum of the form

$$S = \sum_{n=1}^{\infty} a_n.$$

Here a_n are some given real numbers. We would like to have a notion of convergence for series.

Convergence of infinite series We consider the partial sums

$$S_n = \sum_{m=1}^{n} a_m.$$

These are finite sums of numbers. We say that S converges if $\lim_{n \to \infty} S_n$ converges.

If we are given the partial sums S_n, we may recover the terms of the series a_n by

$$a_n = S_n - S_{n-1}.$$

In Section 1.1, we viewed this identity as a form of the fundamental theorem. But, in any case, just as we may convert series to sequences, so we can convert a sequence to a series. We can write

$$\lim_{n \to \infty} b_n = \sum_{n=1}^{\infty} (b_n - b_{n-1}),$$

where we fix b_0 to be zero. Every fact we know about convergence of sequences translates into a fact about convergence of series.

Theorem 2.2.1

The series $\sum_{n=1}^{\infty} a_n$ converges if and only if its tail, $\sum_{n=M}^{\infty} a_n$ converges. (Here M is some particular natural number.)

Proof of Theorem 2.2.1

This is basically just a reformulation of the Cauchy criterion for series. We let S_j be the jth partial sum of the series $\sum_{n=1}^{\infty} a_n$ and we let

$$S_j^M = \sum_{n=M}^{j} a_n.$$

We note that the quantities S_j^M are the partial sums of the tail. Note that if $j,k > M$ then

$$S_j^M - S_k^M = S_j - S_k.$$

We know from Theorems 2.1.1 and 2.1.2 that the tail converges if and only if the S_j^M's are a Cauchy sequence and the original series converges if and only if S_j's are a Cauchy sequence, but restricting N in the definition of Cauchy sequence to numbers greater than M, we see that this is the same thing.

Similarly, we can reformulate the squeeze theorem as a criterion for convergence of series.

Theorem 2.2.2

Let $\{a_n\}$ and $\{b_n\}$ be two sequences of real numbers. Suppose that

$$0 \leq a_n \leq b_n$$

for every natural number n. If $\sum_{n=1}^{\infty} b_n$ converges then $\sum_{n=1}^{\infty} a_n$ converges, and if $\sum_{n=1}^{\infty} a_n$ diverges then $\sum_{n=1}^{\infty} b_n$ diverges.

Proof of Theorem 2.2.2 To address divergence, we apply Theorem 2.1.5 to the partial sums. To address convergence, we observe that the limit of the partial sums is their least upper bound since they are an increasing sequence. Thus our assumption is that the partial sums of the b's have a least upper bound. In particular, since the a's are smaller, this implies that the partial sums of $\sum_{j=1}^{\infty} a_j$ are bounded above. Thus by the least upper bound property of the reals, those partial sums have a least upper bound.

Absolute convergence A series $\sum_{n=1}^{\infty} a_n$ is said to be *absolutely convergent* if $\sum_{n=1}^{\infty} |a_n|$ converges.

Theorem 2.2.3 If $\sum_{n=1}^{\infty} a_n$ converges absolutely, then it converges.

Proof of Theorem 2.2.3 Since $\sum_{n=1}^{\infty} a_n$ is absolutely convergent, it must be that the partial sums of $\sum_{n=1}^{\infty} |a_n|$, which we denote

$$T_n = \sum_{j=1}^{n} |a_j|,$$

are a convergent sequence and therefore a Cauchy sequence. Now denoting by S_n, the nth partial sum of the series $\sum_{n=1}^{\infty} a_n$, we see from the triangle inequality that

$$|S_n - S_m| \leq |T_n - T_m|.$$

Thus $\{S_n\}$ is also a Cauchy sequence and hence converges.

A series $\sum_{n=1}^{\infty} a_n$ need not be absolutely convergent in order to converge.

conditional convergence	If the series converges but is not absolutely convergent, we say that it is conditionally convergent.

Example 1   **This series diverges**

$$\sum_{n=1}^{\infty} (-1)^{n-1} \frac{1}{n}.$$

Proof This sum converges conditionally. To see this, we first observe that the sum does not converge absolutely. This is an application of Example 4 in Section 2.1. Next we combine the $2n-1$st and $2n$th terms of the sum to obtain $\frac{1}{(2n-1)2n}$. The series $\sum_{n=1}^{\infty} (-1)^{n-1} \frac{1}{n}$ converges if and only if

$$\sum_{n=1}^{\infty} \frac{1}{(2n-1)(2n)}$$

converges. We use Theorem 2.2.2 to prove the convergence by comparison with Example 2 of Section 2.1

Example 1 is just one example of a large class of alternating series that converge.

Theorem 2.2.4	Let $\{a_n\}$ be a decreasing sequence of real numbers converging to 0. Then the series $$\sum_{n=1}^{\infty} (-1)^{n-1} a_n$$ converges.

Proof of Theorem 2.2.4   It is enough to show that the series

$$\sum_{n=1}^{\infty} (a_{2n-1} - a_{2n})$$

converges.

Observe that

$$a_{2n-1} - a_{2n} \leq a_{2n-1} - a_{2n+1}.$$

But clearly

$$\sum_{n=1}^{\infty} a_{2n-1} - a_{2n+1} = a_1$$

since it telescopes.

An important example of an absolutely convergent series is the geometric series.

Example 2   Let c and $r < 1$ be positive real numbers. Then

$$\sum_{j=0}^{\infty} cr^j = \frac{c}{1-r}.$$

Proof We can see this by calculating the partial sums

$$S_n = \sum_{j=0}^{n} cr^j = c\left(\frac{1 - r^{n+1}}{1-r}\right).$$

This formula for S_n is most readily seen by induction. It clearly holds for $n = 0$ since the sum is just the 0th term c. We observe that $S_n - S_{n-1} = c(\frac{r^n - r^{n+1}}{1-r}) = cr^n$, which is the nth term. Since $r < 1$, we have that r^{n+1} becomes arbitrarily small as n grows large. Thus S_n converges to $\frac{c}{1-r}$.

We will use the geometric series (Example 2) together with the squeeze theorem for series (Theorem 2.2.2) to devise some useful tests for absolute convergence of series.

Theorem 2.2.5
The ratio test

Suppose $a_n \neq 0$ for any n sufficiently large and suppose that

$$\lim_{n \to \infty} \left| \frac{a_{n+1}}{a_n} \right| = L.$$

If $L < 1$ then the series

$$\sum_{n=1}^{\infty} a_n$$

converges absolutely. If $L > 1$ then the series diverges.

If the limit of the ratios does not exist or is equal to 1, then the ratio test fails, and we can reach no conclusion from Theorem 2.2.5 about the convergence of the series.

Proof of the ratio test

Suppose that $0 \leq L < 1$. Choosing $\epsilon < \frac{1-L}{2}$, we see that there is N, so that for $n \geq N$, we have

$$\left| \frac{a_{n+1}}{a_n} \right| \leq 1 - \epsilon.$$

From this, we see by induction that

$$|a_n| \leq |a_N|(1 - \epsilon)^{n-N}$$

for each $n \geq N$. Now we apply theorem 2.2.1 to see that it suffices to show that the tail of the series,

$$\sum_{n=N}^{\infty} a_n,$$

converges absolutely. To see this, we apply theorem 2.2.2, comparing it with the geometric series

$$\sum_{n=N}^{\infty} |a_N|(1 - \epsilon)^{n-N},$$

Proof of the ratio test cont. which by Example 2 converges absolutely. If, on the other hand, $L > 1$, we may use the same idea to find N and ϵ so that $|a_N| \neq 0$ and so that for $n > N$ we have

$$|a_n| \geq |a_N|(1 + \epsilon)^{n-N}.$$

For such n, it is clear that the differences between consecutive partial sums $|S_{n+1} - S_n| = |a_n|$ are growing. Hence the sequence of partial sums is not a Cauchy sequence.

Theorem 2.2.6 The nth root test Suppose

$$\lim_{n \to \infty} |a_n|^{\frac{1}{n}} = L.$$

Then if $L < 1$, the series $\sum_{n=0}^{\infty} a_n$ converges absolutely. If $L > 1$ then the series diverges.

Proof of The nth root test We proceed just as for Theorem 2.2.5. We suppose $L < 1$ and pick $\epsilon < \frac{1-L}{2}$. Then we conclude that there exists N so that for $n \geq N$ we have

$$|a_n| \leq (1 - \epsilon)^n.$$

Thus, we may apply Theorem 2.2.2 to compare the tail of the series to the tail of the geometric series

$$\sum_{n=N}^{\infty} (1 - \epsilon)^n.$$

On the other hand, if $L > 1$, we see that terms of the series are growing in absolute value, and again we see that the partial sums are not a Cauchy sequence.

The ratio and nth root tests can be used to show that series converge if they do so faster than geometric series. We provide an example.

Example 3   **The series below converges.**

$$\sum_{n=1}^{\infty} n^2 2^{-n}$$

Proof We apply the ratio test and calculate

$$\lim_{n \longrightarrow \infty} \frac{(n+1)^2 2^{-1-n}}{n^2 2^{-n}} = \lim_{n \longrightarrow \infty} \frac{(n+1)^2}{2n^2} = \frac{1}{2}.$$

One of the reasons that the nth root test is important is that we can use it to understand the convergence properties of power series. This will be the topic of our next section.

Exercises for Section 2.2

1. Show using the infinite squeeze theorem that the series

$$\sum_{n=2}^{\infty} \frac{1}{n \log_2 n}$$

diverges. Then show using the squeeze theorem that

$$\sum_{n=2}^{\infty} \frac{1}{n(\log_2 n)^2}$$

converges.

2. Prove either that the following series converges or that it diverges:

$$\sum_{n=1}^{\infty} \frac{n^{2016} 2016^n}{n!}.$$

3. Prove that the following series converges:

$$\sum_{n=1}^{\infty} \frac{n^n}{(n!)^3}.$$

Hint: Find a reasonably good lower bound for $n!$ by a power of n. Don't try to look up the best power of n that's known. Just find a lower bound that you can justify.

4. Prove that

$$\sum_{n=1}^{\infty} \frac{1}{n^r}$$

converges when $r > 1$ and diverges when $r < 1$. Hint: Break up the sum into dyadic pieces; that is $2^j \le n \le 2^{j+1}$. Bound the sums of each piece above when $r > 1$ and below when $r < 1$. Note: The function x^r hasn't actually been defined yet, but you may use all its basic properties, like $x^r = x(x^{r-1})$ and that positive powers of numbers greater than 1 are greater than 1.

5. Prove that

$$\sum_{n=1}^{\infty} 2^{\sqrt{n}-n}$$

converges.

6. A thirty-year fixed-rate mort-
   gate is a loan taken out over
   a period of 360 months. The
   initial loan amount is M.
   Each month, the borrower
   pays a fixed payment p. We
   define a function $f(j)$ where j
   is the number of months that
   have passed. We let $f(0) =
   M$ and we let $f(j) = (1 +
   r)f(j-1) - p$ for $1 \leq j \leq 360$,
   where r is the fixed monthly
   interest rate. Further, we require that $f(360) = 0$. Derive and prove a formula for
   p in terms of M and r in
   closed form.   Hint: You'll
   have to use the formula for the
   sum of a finite geometric series, which appears in thep-
   roof of Example 2. It helps
   to rearrange things so that
   you're setting the mortgage
   amount M with interest on it
   compounded over thirty years
   equal the stream of monthly
   payments each compounded
   from the moment it is made.
   Aside: This is really how
   payments on thirty-year fixed-
   rate mortgages are computed.

Flipped recitation, ratio and root tests

In this section, we studied the ratio test and nth root test for convergence of series.

Problem 1

Show using the ratio test that

$$\sum_{n=1}^{\infty} n^{100} 3^{-n}$$

converges

Problem 2

Show using the nth root test that

$$\sum_{n=1}^{\infty} n^{200} 5^n$$

diverges.

Problem 3

Show that neither the ratio test nor the nth root test give any result by direct application to the series

$$\sum_{n=1}^{\infty} n^{1000}$$

or

$$\sum_{n=1}^{\infty} n^{-1000}.$$

◇ 2.3   Power series

A very important class of series to study is the power series. Power series are interesting in part because they represent functions and in part because they encode their coefficients, which are a sequence. At the end of this section, we will see an application of power series for writing a formula for an interesting sequence.

Power series   A power series is an expression of the form

$$S(x) = \sum_{j=0}^{\infty} a_j x^j.$$

For the moment, the coefficients a_j will be real numbers. The variable x takes real values, and for each distinct value of x we get a different series $S(x)$. The first question we'll be interested in is for what values of x the series $S(x)$ converges.

Theorem 2.3.1   Let

$$S(x) = \sum_{j=0}^{\infty} a_j x^j.$$

Then there is a unique $R \in [0,\infty]$ so that $S(x)$ converges absolutely when $|x| < R$ and so that $S(x)$ diverges when $|x| > R$.

Radius of convergence   The number R (possibly infinite) which Theorem 2.3.1 guarantees is called the *radius of convergence* of the power series.

Often, to prove a theorem, we break it down into simpler parts which we call lemmas. This is going to be one of those times.

Lemma 1

Let

$$S(x) = \sum_{j=0}^{\infty} a_j x^j.$$

Suppose that $S(c)$ converges. Then $S(x)$ converges absolutely for all x so that $|x| < |c|$.

Proof of Lemma 1

We note that since $S(c)$ converges, it must be that the sequence of numbers $\{|a_j c^j|\}$ is bounded above. If not, there are arbitrarily late partial sums of $S(c)$ which differ by an arbitrarily large quantity, implying that the series $S(c)$ does not converge. Let K be an upper bound for the sequence $\{|a_j c^j|\}$. Now suppose $|x| < |c|$. We will show that $S(x)$ converges absolutely. Observe that we have the inequality

$$|a_j x^j| \leq K|(\frac{x}{c})|^j.$$

By Theorem 2.2.2 , it suffices to show that the series

$$\sum_{j=0}^{\infty} K|(\frac{x}{c})|^j$$

converges. But this is true since the series above is geometric and by assumption $|\frac{x}{c}| < 1$.

Now we are in a strong position to prove Theorem 2.3.1.

Proof of   We will prove Theorem 2.3.1 by defining R.
Theorem 2.3.1 We let R be the least upper bound of the set of
$|x|$ so that $S(x)$ converges. If this set happens
not to be bounded above, we let $R = \infty$. By
the definition of R, it must be that for any
x with $|x| > R$, we have that $S(x)$ diverges.
(Otherwise R isn't an upper bound.) Now
suppose that $|x| < R$. Then there is y with
$|y| > |x|$ so that $S(y)$ converges. (Otherwise,
$|x|$ is an upper bound.) Now, we just apply
Lemma 1 to conclude that $S(x)$ converges.

The above proof gives the radius of convergence R in terms
of the set of x where the series converges. We can, however, determine R in terms of the coefficients of the series. We consider
the sets

$$A_k = \{|a_n|^{\frac{1}{n}} : n \geq k\}.$$

These are the sets of nth roots of nth coefficients in the tail
of the series. Let T_k be the least upper bound of A_k. The
numbers T_k are a decreasing sequence of positive numbers and
have a limit unless they are all infinite. Let

$$T = \lim_{k \longrightarrow \infty} T_k.$$

Then T is either a nonnegative real number or is infinite. It
turns out that $R = \frac{1}{T}$. You are asked to show this in an exercise,
but it is a rather simple application of the nth root test. This
is the reason the nth root test is important for understanding
power series.

One thing we haven't discussed yet is the convergence of the
power series right at the radius of convergence. Basically, all
outcomes are possible. Directly at the radius of convergence,
we are in a setting where the nth root test fails.

Example 1   Consider the following three series.

$$S_1(x) = \sum_{n=0}^{\infty} x^n.$$

$$S_2(x) = \sum_{n=1}^{\infty} \frac{x^n}{n}.$$

$$S_3(x) = \sum_{n=1}^{\infty} \frac{x^n}{n^2}.$$

By the criterion above, it is rather easy to see that the radius of convergence of each series is 1, since the nth roots of the coefficients converge to 1. However, the three series have rather different behaviors at the points $x = 1$ and $x = -1$. We note that $S_1(x)$ diverges at both $x = 1$ and $x = -1$ since all of its terms there have absolute value 1. We note that $S_2(1)$ is the harmonic series, which diverges, and we note that $S_2(-1)$ is the alternating version of the harmonic series, which we showed converges conditionally. We can see that $S_3(1)$ and $S_3(-1)$ both converge absolutely since they can be compared with the series

$$\sum_{n=1}^{\infty} \frac{1}{n^2}.$$

Since we are interested in studying power series as functions and we are accustomed to adding and multiplying functions, it will be important to us to understand that we can add and multiply absolutely convergent series termwise. Once we have done this, we will see that we can do the same with power series inside their radii of convergence.

Theorem 2.3.2

Let $S_1 = \sum_{n=0}^{\infty} a_n$ and $S_2 = \sum_{n=0}^{\infty} b_n$ be absolutely convergent series. Then

$$S_1 + S_2 = \sum_{n=0}^{\infty} a_n + b_n,$$

and letting

$$c_m = \sum_{i+j=m} a_i b_j,$$

we have

$$S_1 S_2 = \sum_{n=0}^{\infty} c_n.$$

It is worth noting that even the statement of the Theorem 2.3.2 for products looks a little more complicated than the one for sums. The issue is that products of partial sums are not exactly the partial sums of the products.

Proof of Theorem 2.3.2

The proof of the statement about sums is essentially immediate since the partial sums of the formula for sums are the sums of partial sums of the individual series. So we need only check that the limit of a sum is the sum of the limits, which we leave to the reader. For products, things are a little more complicated.

Proof of Theorem 2.3.2 cont. We observe that the sum of an absolutely convergent series is the difference between the sum of the series of its positive terms and the sum of the series of its negative terms, and so we restrict our attention to the case where all a_i's and all b_i's are nonnegative. We let $S_{1,n}$ be the nth partial sum of S_1, we let $S_{2,n}$ be the nth partial sum of S_2 and we let $S_{3,n}$ be the nth partial sum of

$$\sum_{n=0}^{\infty} c_n.$$

Then we notice that

$$S_{1,n}S_{2,n} \leq S_{3,2n} \leq S_{1,2n}S_{2,2n}.$$

We obtain the desired conclusion using the squeeze theorem.

Theorem 2.3.2 guarantees us that we can multiply two power series formally if we are inside both of their radii of convergence. If

$$S_1(x) = \sum_{j=0}^{\infty} a_j x^j,$$

and

$$S_2(x) = \sum_{j=0}^{\infty} b_j x^j,$$

then with

$$c_l = \sum_{j+k=l} a_j b_k x^l$$

we have that

$$\sum_{l=0}^{\infty} c_l$$

is a power series with c_l the lth term.

Later, when we study Taylor's theorem, we will establish power series expressions for essentially all the functions that we know how to differentiate. As it is, we already know power series expansions for a large class of functions because of our familiarity with geometric series.

Example 2

$$\frac{1}{1-ax} = \sum_{n=0}^{\infty} (ax)^n$$

whenever $|ax| < 1$.

Proof The equality expressed above is just a special case of the formula for the sum of an infinite geometric series. However the right-hand side is a power series expression for the function on the left-hand side. The radius of convergence of the series is $\frac{1}{|a|}$, which is the distance from zero to the singularity of the function.

In conjunction with Theorem 2.3.2, we can actually use this formula to obtain the power series at zero of any rational function. Suppose

$$f(x) = \frac{P(x)}{Q(x)}$$

is a rational function (That is, $P(x)$ and $Q(x)$ are polynomials.) Suppose moreover that the roots of $Q(x)$ are distinct. Let us call them $r_1, \ldots, r_m$. Then by partial fractions decomposition

$$f(x) = S(x) + \frac{A_1}{x - r_1} + \cdots + \frac{A_m}{x - r_m},$$

where $S(x)$ is a polynomial and the A's are constants. Using geometric series, we already have a series expansion for each term in this sum.

What happens if $Q(x)$ does not have distinct roots? Then we need power series expansions for $\frac{1}{(x-r)^2}, \frac{1}{(x-r)^3}, \ldots$. Later, we'll see that an easy way of getting them is by differentiating the series for $\frac{1}{x-r}$. But as it is, we can also get the series by taking $\frac{1}{x-r}$ to powers. For instance,

$$\frac{1}{(1-ax)^2} = (\sum_{n=0}^{\infty} (ax)^n)^2 = \sum_{n=0}^{\infty} (n+1)(ax)^n.$$

·Here what we have done is simply apply the multiplication part of Theorem 2.3.2. As long as we can count the number of terms in the product, we can obtain a series expansion for any rational function.

Example 3   The Fibonacci sequence

As promised, we will now use the theory of power series to understand the terms of an individual sequence. We now define the Fibonacci sequence. We begin by letting $f_0 = 1$ and $f_1 = 1$. Then for $j \geq 2$, we let

$$f_j = f_{j-1} + f_{j-2}.$$

The above formula is called the recurrence relation for the Fibonacci sequence and it lets us generate this sequence one term at a time:

$$f_0 = 1, f_1 = 1, f_2 = 2, f_3 = 3, f_4 = 5, f_5 = 8, f_6 = 13, f_7 = 21, \ldots$$

The Fibonacci sequence is much loved by math geeks and has a long history. It was used by Leonardo Fibonacci in the thirteenth century to model populations of rabbits for reasons that are too upsetting to relate.

Nevertheless, our present description of the sequence is disturbingly inexplici. To get each term, we need first to have computed the previous two terms. This situation is sufficiently alarming that one well-known calculus book gives the Fibonacci sequence as an example of a sequence whose nth term cannot be described by a simple formula. Using power series, we are now in a position to make a liar of that calculus book.

We introduce the following power series

$$f(x) = \sum_{n=0}^{\infty} f_n x^n,$$

which has the Fibonacci sequence as its coefficients. We note that multiplying $f(x)$ by a power of x shifts the sequence. We consider the expression $(1 - x - x^2)f(x)$ and note that by the recurrence relation, all terms with x^2 or higher vanish. Computing the first two terms by hand, we see that

$$(1 - x - x^2)f(x) = 1,$$

or, put differently,

$$f(x) = \frac{1}{1 - x - x^2}.$$

Applying partial fractions, we conclude that

$$f(x) = \frac{-\frac{1}{\sqrt{5}}}{x + \frac{1+\sqrt{5}}{2}} + \frac{\frac{1}{\sqrt{5}}}{x + \frac{1-\sqrt{5}}{2}}.$$

Now applying the formula for the sum of a geometric series and using the fact that

$$(\frac{1 + \sqrt{5}}{2})(\frac{1 - \sqrt{5}}{2}) = -1,$$

we see that

$$f_n = \frac{1}{\sqrt{5}}(\frac{1 + \sqrt{5}}{2})^n - \frac{1}{\sqrt{5}}(\frac{1 - \sqrt{5}}{2})^n.$$

What could be simpler?

Exercises for Section 2.3

1. Show that

$$\sum_{n=1}^{\infty} n^n x^n$$

   diverges for all $x > 0$.

2. Find the radius of convergence of

$$\sum_{n=1}^{\infty} \sqrt{n} 4^n x^n.$$

   Justify your answer, of course.

3. Let $\{a_n\}$ be a sequence satisfying $a_n = 2a_{n-1} + 3a_{n-2}$ for $n > 2$ with $a_1 = 1$ and $a_2 = 2$. Following Example 3, find a rational function representing the power series

$$\sum_{n=1}^{\infty} a_n x^n.$$

   What is the radius of convergence of this series? Justify your answer. Hint: The sequence a_n is a sum of two geometric sequences.

4. Let a_n be a sequence of real numbers bounded above and below. For each n, let b_n be the least upper bound of

$$\{a_k : k > n\}.$$

   Prove that b_n is a decreasing sequence. Define

$$\limsup a_n$$

   to be the greatest lower bound of b_n. (That is, $\limsup a_n$ is the negative of the least upper bound of $\{-b_n\}$.) Prove there is a subsequence of $\{a_n\}$ which converges to $\limsup a_n$. Hint: This is just going through the definition and finding lots of a's close to $\limsup a_n$.

5. With a_n and $\limsup a_n$ as in the previous exercise, let L be the limit of some subsequence of a_n. Show that $L \leq \limsup a_n$. Hint: Compare L to the b_n's.

6. Let $\{a_n\}$ be a positive sequence of real numbers. Suppose that

$$L = \limsup a_n^{\frac{1}{n}}$$

is nonzero and finite. Show that $\frac{1}{L}$ is the radius of convergence of the power series

$$\sum_{n=0}^{\infty} a_n x^n.$$

Hint: There are two parts to this exercise. You need an upper bound and a lower bound for the radius of convergence. To get the lower bound just use the nth root test. To get the upper bound, use the subsequence which converges to the $\limsup$.

Flipped recitation, power series, radius of convergence

In the section, we studied absolute convergence of power series and discovered that except in boundary cases, it is largely governed by a single parameter, the radius of convergence.

Problem 1

Consider the power series

$$f(x) = \sum_{n=0}^{\infty} n! x^n.$$

What is the radius of convergence?

Problem 2

Consider the power series

$$f(x) = \sum_{n=0}^{\infty} \frac{x^n}{n!}.$$

What is the radius of convergence?

Problem 3

Let a and b be positive real numbers. What is a power series for the function

$$f(x) = \frac{1}{(x-a)(x-b)}?$$

What is the radius of convergence for this power series?

Chapter 3

FUNCTIONS AND DERIVATIVES

◇ 3.1 Continuity and limits

In this section, we'll be discussing limits of functions on the reals, and for this reason we have to modify our definition of limit. For the record:

Functions

> A function f from the reals to the reals is a set G of ordered pairs (x,y) so that for any real number x, there is at most one y with $(x,y) \in G$. The set of x for which there is a y with $(x,y) \in G$ is called the domain of the function. If x is in the domain, the real number y for which $(x,y) \in G$ is called $f(x)$.

Don't panic! I don't blame you if the above definition, beloved of mathematicians, is not how you usually think of functions. The set G is usually referred to as the graph of the function. The condition that there is only one y for each x is the vertical line test. However, all of this is still a little drier than the way we usually imagine functions. We like to think there is a formula, a rule, which tells us how we compute $f(x)$ given x. Certainly some of our favorite functions arise in that way, but it is not the case that most functions do, even granting some ambiguity in what we mean by a formula or a rule. Nonetheless, in this section we will deal with functions at this level of generality. One consolation might be that when you are out in nature collecting data to determine a function, your data will come as points of its graph (or rather approximations to them since in reality we don't see real numbers.)

Limits of functions

If f is a function on the reals, [or possibly on the reals excluding the point a], we say that

$$\lim_{x \to a} f(x) = L$$

if for every $\epsilon > 0$, there is $\delta > 0$ so that if $0 < |x - a| < \delta$, then $|f(x) - L| < \epsilon$.

The definition of the limit should by now look somewhat familiar. Because we are looking at limits of a function instead of limits of a sequence, the quantity $N(\epsilon)$, which measured how far in the sequence we had to go to get close to the limit, is replaced by the quantity $\delta(\epsilon)$, which measures how close we have to be to a for the function f to be close to its limit.

To get a handle on how a definition works, it helps to work through some examples.

Example 1   Show that

$$\lim_{h \to 0} \frac{(2 + h)^2 - 4}{h} = 4.$$

Proof Here the function $f(h) = \frac{(2+h)^2-4}{h}$ is technically not defined at 0. However, at every other h we see that the function is the same as $4 + h$. Hence the problem is the same as showing

$$\lim_{h \to 0} 4 + h = 4.$$

Thus what we need to do is find a $\delta(\epsilon)$ so that $|4+h-4| < \epsilon$, when $|h| < \delta(\epsilon)$. Since $|4 + h - 4|$ is the same as $|h|$, we just use $\delta(\epsilon) = \epsilon$.

A lot of the limits we can take in elementary calculus work as in Example 1. We rewrite the function whose limit we are taking on its domain in a way that makes it easier for us to estimate the difference between the function and its limit.

The rules that we had for taking limits of sequences still work for limits of functions.

Theorem 3.1.1 squeeze theorem for functions

Let f, g, h be functions which are defined on the reals without the point a. Suppose that everywhere we know that $f(x) \leq h(x) \leq g(x)$ and suppose that

$$\lim_{x \to a} f(x) = \lim_{x \to a} g(x) = L.$$

Then

$$\lim_{x \to a} h(x) = L.$$

The proof basically repeats the proof of the squeeze theorem for sequences.

Proof of the squeeze theorem for functions

We can find a common function $\delta(\epsilon)$ so that $|x - a| < \delta(\epsilon)$ implies that $|f(x) - L| < \epsilon$ and $|g(x) - L| < \epsilon$. Then we observe that $f(x) - L \leq h(x) - L \leq g(x) - L$. Thus

$$|h(x) - L| \leq \max(|f(x) - L|, |g(x) - L|) < \epsilon,$$

where the last inequality only holds when $|x - a| < \delta(\epsilon)$. Thus we have shown that

$$\lim_{x \to a} h(x) = L.$$

The notion of limit allows us to introduce the notion of a continuous function. We first write down a helpful lemma

Lemma 3.1.1   Let $\{x_j\}$ be a sequence of real numbers converging to a. Suppose that

$$\lim_{x \to a} f(x) = L.$$

Then

$$\lim_{j \to \infty} f(x_j) = L.$$

Proof of Lemma 3.1.1   We need to show that for every $\epsilon > 0$, there is $N(\epsilon)$ so that if $n > N(\epsilon)$, then $|L - f(x_j)| < \epsilon$. What we do know is that for every $\epsilon > 0$ there is $\delta > 0$ so that if $|x - a| < \delta$, then $|f(x) - L| < \epsilon$. Thus it would be enough to show that there is N so that if $j > N$ then $|x_j - a| < \delta$. This we know from the convergence of the sequence to a, using δ in the role of ϵ.

Continuous function   A function f on the reals is continuous at a point a if

$$\lim_{x \to a} f(x) = f(a).$$

We say that f is continous on an interval $[c,d]$ if it is continuous for every $a \in [c,d]$

We shall now take some time to prove as theorems some of the basic properties of continuous functions that we tend to take for granted. In the following theorem, we shall use l.u.b. to denote least upper bound.

Theorem 3.1.2 extreme value theorem

Let $f(x)$ be a function which is continuous on the interval $[a,b]$. Then $f(x)$ attains its maximum on this interval. More precisely, if $M = l.u.b.\{f(x) : x \in [a,b]\}$, then M exists and there is a point $c \in [a,b]$ so that

$$f(c) = M.$$

Proof of extreme value theorem

The hardest part of proving this theorem is to show that the set $\{f(x) : x \in [a,b]\}$, which is clearly nonempty, is bounded above. We prove this by contradiction. Suppose the set is not bounded above. Then for every natural number n, there is $x_n \in [a,b]$ so that $f(x_n) > n$. (Otherwise n is an upper bound.) Now we apply the Bolzano-Weierstrass theorem. This tells us that there is a subsequence x_{n_j} converging to some point $z \in [a,b]$. But by the definition of continuity

$$\lim_{j \to \infty} f(x_{n_j}) = f(z) < \infty,$$

which is impossible since by assumption $f(x_{n_j}) > n_j$.

Now we know that $\{f(x) : x \in [a,b]\}$ is bounded above. Therefore it has a least upper bound. Let us denote the least upper bound by M. Since M is the least upper bound, it is the case that for every n, there is a point $x_n \in [a,b]$ so that

$$M - \frac{1}{10^n} < f(x_n) \leq M.$$

(Otherwise $M - \frac{1}{10^n}$ is also an upper bound and so M is not the least upper bound.)

Proof of extreme value theorem cont. Now applying the Bolzano-Weierstrass theorem again, we see that there is a subsequence $\{x_{n_j}\}$ converging to some point $c \in [a,b]$. By the definition of continuity, we have that

$$\lim_{j \to \infty} f(x_{n_j}) = f(c).$$

Thus we see that

$$f(c) = M.$$

Since M is the least upper bound, we see that f achieves a maximum at c.

The key ingredient in the proof of the extreme value theorem was the Bolzano-Weierstrass theorem. It was there that we used seriously the important hypothesis that the domain on which the function is continuous is a closed interval.

We are now ready to prove the other most iconic property of continuous functions:

Theorem 3.1.3 intermediate value theorem Let f be a continuous function on the interval $[a,b]$. Suppose that $f(a) < L < f(b)$. Then there is some $c \in [a,b]$ so that $f(c) = L$.

Proof of intermediate value theorem We will prove this theorem by contradiction. Suppose there is no value c for which $f(c) = L$. We consider the midpoint of the interval $\frac{a+b}{2}$. By assumption, either $f(\frac{a+b}{2}) < L$ or $f(\frac{a+b}{2}) > L$.

Proof of intermediate value theorem cont.

If $f(\frac{a+b}{2}) < L$, we define new endpoints $a_1 = \frac{a+b}{2}$ and $b_1 = b$. If $f(\frac{a+b}{2}) > L$, we define instead $a_1 = a$ and $b_1 = \frac{a+b}{2}$. In either case, we have that the hypotheses of the theorem are retained with a replaced by a_1 and b replaced by b_1. Moreover, we have that each of the three numbers $a_1 - a$, $b - b_1$, and $b_1 - a_1$ is bounded by $\frac{b-a}{2}$.

We keep repeating this process, shrinking the interval by a factor of two each time. Thus we obtain sequences $\{a_l\}$ and $\{b_l\}$ so that $f(a_l) < L$, so that $f(b_l) > L$, and so that the three numbers $a_l - a_{l-1}$, $b_{l-1} - b_l$, and $a_l - b_l$ are all nonnegative and bounded above by $\frac{b_{l-1}-a_{l-1}}{2} = \frac{b-a}{2^l}$. Thus we have that $\{a_l\}$ and $\{b_l\}$ are Cauchy sequences converging to the same point c. Thus, by the definition of continuity, the sequences $\{f(a_l)\}$ and $\{f(b_l)\}$ both converge to the same limit $f(c)$. But since for all L we have

$$f(a_l) < L < f(b_l),$$

by the squeeze theorem we have that $f(c) = L$. This is a contradiction.

We close the section with a useful result for evaluating limits (in this case of sequences.) The same result is true for limits of functions, as you'll see in the exercises.

Theorem 3.1.4
Limit laws

Let $\{a_n\}$ and $\{b_n\}$ be sequences. Suppose that

$$\lim_{n \to \infty} a_n = L_1$$

and

$$\lim_{n \to \infty} b_n = L_2.$$

Then

$$\lim_{n \to \infty} a_n + b_n = L_1 + L_2.$$

Moreover,

$$\lim_{n \to \infty} a_n b_n = L_1 L_2.$$

Proof of Theorem 3.1.4

We begin by proving

$$\lim_{n \to \infty} a_n + b_n = L_1 + L_2.$$

We observe that there is $N_1 > 0$ so that when $n > N_1$, we have

$$|a_n - L_1| < \frac{\epsilon}{2},$$

and that there is $N_2 > 0$, so that when $n > N_2$,

$$|b_n - L_2| < \frac{\epsilon}{2}.$$

Thus, letting N be the larger of N_1 and N_2, we get that when $n > N$,

$$|a_n + b_n - L_1 - L_2| \leq |a_n - L_1| + |b_n - L_2| < \epsilon.$$

**Proof of
Theorem
3.1.4 cont.**

To prove the same result for products is a bit more complicated. We calculate

$$|L_1 L_2 - a_n b_n|$$

$$\leq |L_1 L_2 - L_1 b_n| + |L_1 b_n - a_n b_n|$$

$$\leq |L_1||L_2 - b_n| + |b_n||L_1 - a_n|.$$

We observe that since b_n converges, it must be that b_n is a bounded sequence, and we let M be the least upper bound of $\{|b_n|\}$. Thus our estimate becomes

$$|L_1 L_2 - a_n b_n| \leq |L_1||L_2 - b_n| + M|L_1 - a_n|.$$

Now we use the fact that $\{a_n\}$ converges to L_1 and $\{b_n\}$ to L_2. There is N_1 so that for $n > N_1$, we have

$$|L_1 - a_n| < \frac{\epsilon}{2M}.$$

There is N_2 so that for $n > N_2$,

$$|L_2 - b_n| < \frac{\epsilon}{2L_1}.$$

Then

$$|L_1 L_2 - a_n b_n| < \epsilon.$$

Exercises for Section 3.1

1. Prove the limit laws for functions. That is, suppose that

$$\lim_{x \to a} f(x) = L_1$$

and

$$\lim_{x \to a} g(x) = L_2.$$

Show that

$$\lim_{x \to a} f(x) + g(x) = L_1 + L_2$$

and that

$$\lim_{x \to a} f(x)g(x) = L_1 L_2.$$

2. Show that the function $f(x) = \sqrt{x}$ is continuous at $x = 1$.

3. Let $f(x) = x^3 + ax^2 + bx + c$ be a cubic polynomial with coefficients a, b, c real. Show that there is some real y so that $f(y) = 0$.

4. Prove that

$$\lim_{h \to 0} \frac{\sqrt{1 + 3h} - 1}{h} = \frac{3}{2}.$$

Hint: Multiply the numerator and the denominator by $\sqrt{1 + 3h} + 1$.

5. We say that a function f is *uniformly continuous* on an interval $[c,d]$ if for every $\epsilon > 0$ there exists $\delta(\epsilon) > 0$ so that if $x, y \in [c,d]$ with $|x - y| < \delta(\epsilon)$, we have $|f(x) - f(y)| < \epsilon$. Note that this is stronger than the definition of continuity because $\delta(\epsilon)$ does not depend on the point of continuity. Show that any function continuous on all of $[c,d]$ is uniformly continuous on $[c,d]$. Hint: By continuity, at each point of the interval $[c,d]$, there is a $\delta(\epsilon)$ appropriate for that point. If these numbers are bounded below by something greater than 0, take the lower bound. Otherwise, use the Bolzano-Weierstrass theorem to find a point of discontinuity, obtaining a contradiction.

Flipped recitation, Limits of functions
In this section, we introduced limits of functions and continuity of functions, and we proved the extreme value theorem and intermediate value theorem.

Problem 1
Show that the polynomial $f(x) = x^4 - 4x + 1$ has a root in the interval $[1,2]$.

Problem 2
Calculate
$$\lim_{x \to 2} \frac{x^2 - 3x + 2}{x^2 - 5x + 6}.$$

Problem 3
Show that the function $f(x) = x^{\frac{1}{3}}$ is continuous at $x = 8$. Hint: Note that $f(8) = 2$. Consider $(2 \pm \epsilon)^3$. Convert appropriate bounds to good choices of δ.

◇ 3.2 Derivatives

In this section, we will define the derivative of a function and describe its familiar local theory.

Before doing this we will introduce a bit of notation, common in applied fields like the analysis of algorithms, but not often used when discussing single variable calculus. We will do so because it makes the proofs of the main rules of differentiation, like the product and the chain rules, extremely transparent.

Little-oh and big-oh notation

We say that a function $f(h)$ is $o(h)$ if as $h \longrightarrow 0$,

$$\lim_{h \longrightarrow 0} \frac{f(h)}{h} = 0.$$

More generally, if $g(h)$ is a continuous increasing function of h with $g(0) = 0$, we say that $f(h)$ is $o(g(h))$ if

$$\lim_{h \longrightarrow 0} \frac{f(h)}{g(|h|)} = 0.$$

We say that f is $O(h)$ as $h \longrightarrow 0$ if there exist $C, \epsilon > 0$ so that for $|h| < \epsilon$, we have

$$|f(h)| \leq C|h|.$$

More generally, if $g(h)$ is a continuous increasing function of h with $g(0) = 0$, we say that $f(h)$ is $O(g(h))$ if there exist $C, \epsilon > 0$ so that for $|h| < \epsilon$, we have

$$|f(h)| \leq Cg(|h|).$$

Big-oh and little-oh notation are about describing the size of functions of h asymptotically as $h \longrightarrow 0$. The reason the letter "oh" is used here is that it is the first letter of the word "order". When f is $O(g)$, this means that f is roughly the same size as g. [Here "roughly" means neglecting the factor of C and only considering h near 0.] We often say for this that f is of the order of g. If f is $o(g)$, it means that f is much smaller than g.

Now we will see how this relates to differentiation. Let us first give the familiar definition of the derivative.

Derivative of a function A function f is differentiable at x if

$$\lim_{h \longrightarrow 0} \frac{f(x+h) - f(x)}{h}$$

exists. We denote this limit by $f'(x)$ or $\frac{d}{dx}(f(x))$.

We can reformulate this using little-oh notation. A function f is differentiable at x if and only if there is a number $f'(x)$ so that

$$f(x+h) = f(x) + hf'(x) + o(h).$$

(Here $+o(h)$ is shorthand for adding a function which is $o(h)$.) The formula above is called the differential approximation for f. It says that if we ignore an $o(h)$ error, the function f is approximated by a linear function with slope $f'(x)$.

Our philosophy throughout will be that we should always think of differentiability in terms of the differential approximation. Almost without fail, if you are trying to prove a theorem and you are given in the hypotheses that a function is differentiable, you should consider applying the differential approximation.

We see immediately from the differential approximation that if f is differentiable at x, then

$$f(x+h) = f(x) + O(h).$$

From this it can be shown that if f is differentiable at x then f is continuous at x.

Theorem 3.2.1
The product rule

If $f(x)$ and $g(x)$ are functions differentiable at x, then the product $f(x)g(x)$ is differentiable at x and

$$\frac{d}{dx}(f(x)g(x)) = f(x)g'(x) + g(x)f'(x).$$

Proof of the product rule

Since f and g are differentiable at x, we have

$$f(x + h) = f(x) + hf'(x) + o(h)$$

and

$$g(x + h) = g(x) + hg'(x) + o(h).$$

Now we multiply these two equations together.

$$f(x + h)g(x + h)$$
$$= (f(x) + hf'(x) + o(h))(g(x) + hg'(x) + o(h))$$
$$= f(x)g(x) + h(f(x)g'(x) + g(x)f'(x)) + o(h).$$

Thus the theorem is proved.

Implicitly what we used in this proof is that h^2 is $o(h)$ and that anything which is $O(1)$ multiplied by something $o(h)$ is $o(h)$. These multiplication rules are made more explicit in the exercises.

Theorem 3.2.2
The chain rule

Suppose that g is differentiable at x and f is differentiable at $g(x)$. Then the function $q(x) = f(g(x))$ is differentiable at x and

$$q'(x) = f'(g(x))g'(x).$$

Proof of the Chain rule

We calculate

$$q(x + h) - q(x) \tag{3.1}$$
$$= f(g(x + h)) - f(g(x)) \tag{3.2}$$
$$= [g(x+h)-g(x)]f'(g(x))+o(g(x+h)-g(x)). \tag{3.3}$$

Here, in (3.3), we have just used the differentiability of f at $g(x)$. Now since g is differentiable at x, we have that $g(x + h) - g(x)$ is $O(h)$.

In (3.3), we have $o(g(x + h) - g(x))$, which is $o(O(h))$, which is $o(h)$.

Thus, rewriting the equation, we get

$$q(x + h) - q(x) \tag{3.4}$$
$$= [g(x + h) - g(x)]f'(g(x)) + o(h) \tag{3.5}$$
$$= [g'(x)h + o(h)]f'(g(x)) + o(h) \tag{3.6}$$
$$= f'(g(x))g'(x)h + o(h). \tag{3.7}$$

Here in (3.6), we have used the differentiability of g at x. Thus we have proved the theorem.

We can go a long way towards building up all of differential calculus using just the product rule and the chain rule (as well as some simpler things like the sum rule).

Theorem 3.2.3
The sum rule

If $f(x)$ and $g(x)$ are differentiable at x, then

$$\frac{d}{dx}(f(x) + g(x)) = f'(x) + g'(x).$$

Proof of the sum rule

$$(f(x+h) + g(x+h) - f(x) - g(x))$$
$$= h(f'(x) + g'(x)) + o(h).$$

Here we have used the commutativity of addition as well as the fact that $o(h) + o(h) = o(h)$.

Theorem 3.2.4
power rule for the natural numbers

Let $n \in \mathbf{N}$. Let $f(x) = x^n$. Then

$$f'(x) = nx^{n-1}.$$

Proof of the power rule for natural numbers

Of course, we prove this by induction on n. First we do the base case.

$$(x+h) - x = h = h + o(h).$$

Here we used the fact that 0 is $o(h)$.
Now we do the induction step. We assume that the derivative of x^{n-1} is $n - 1x^{n-2}$. We write

$$f(x) = x^{n-1}x.$$

Now we apply the product rule.

$$f'(x) = (n-1)x^{n-2}x + x^{n-1} = nx^{n-1}.$$

Theorem 3.2.5
Quotient rule, Version 1

Suppose $f(x), g(x)$ and $\frac{f(x)}{g(x)}$ are differentiable at x and $g(x) \neq 0$. Then

$$\frac{d}{dx}\left(\frac{f(x)}{g(x)}\right) = \frac{f'(x)g(x) - f(x)g'(x)}{(g(x))^2}.$$

Proof of quotient rule, Version 1

We just write

$$f(x) = \left(\frac{f(x)}{g(x)}\right)g(x)$$

and apply the product rule, getting

$$f'(x) = \frac{d}{dx}\left(\frac{f(x)}{g(x)}\right)g(x) + g'(x)\left(\frac{f(x)}{g(x)}\right).$$

We now solve for $\frac{d}{dx}\left(\frac{f(x)}{g(x)}\right)$.

The quotient rule is in fact a bit stronger. We do not need to assume *a priori* that the quotient is differentiable.

Theorem 3.2.6 Quotient rule, Version 2)

Let $f(x)$ and $g(x)$ be differentiable at x and let $g(x) \neq 0$. Then $\frac{f(x)}{g(x)}$ is differentiable.

Proof of quotient rule, Version 2

$$\frac{f(x+h)}{g(x+h)} - \frac{f(x)}{g(x)} \tag{3.8}$$

$$= \frac{f(x+h)g(x) - f(x)g(x+h)}{g(x)g(x+h)} \tag{3.9}$$

$$= \frac{f(x)g(x) - f(x)g(x) + (f'(x)g(x) - f(x)g'(x))h + o(h)}{g(x)g(x+h)} \tag{3.10}$$

$$= \frac{(f'(x)g(x) - f(x)g'(x))h + o(h)}{g(x)(g(x) + O(h))} \tag{3.11}$$

$$= \frac{f'(x)g(x) - f(x)g'(x)}{(g(x))^2}h + o(h). \tag{3.12}$$

We can use the chain rule to obtain the inverse rule.

Theorem 3.2.7 Inverse rule, version 1   Suppose that $f(g(x)) = x$ and g is differentiable at x and f is differentiable at $g(x)$ then

$$f'(g(x)) = \frac{1}{g'(x)}.$$

Proof of inverse rule, version 1   We just apply the chain rule to $f(g(x)) = x$ and solve for $f'(g(x))$.

In fact, there is a stronger version of the inverse rule, guaranteeing the differentiability of the inverse f at $g(x)$ if $g'(x) \neq 0$ and in fact guaranteeing that f exists under that condition. We will discuss that in a later section.

An application is that this allows us to differentiate rational powers. We define

$$x^{\frac{1}{n}} = l.u.b\{y : y^n < x\}.$$

It is easy to see that

$$(x^{\frac{1}{n}})^n = x.$$

Differentiating both sides, using the chain rule on the left side, we get

$$1 = \frac{d}{dx}(x^{\frac{1}{n}})^n \tag{3.13}$$

$$= n(x^{\frac{1}{n}})^{n-1}\frac{d}{dx}(x^{\frac{1}{n}}). \tag{3.14}$$

We solve obtaining

$$\frac{d}{dx}(x^{\frac{1}{n}}) = \frac{1}{n}x^{-\frac{n-1}{n}}.$$

We can actually define irrational powers as limits of rationals but we delay this to our study of exponential functions. As it is, we can differentiate all algebraic functions, and this is a course in which we do "late transcendentals."

Exercises for Section 3.2

1. Let $f(h)$ be $O(1)$ as $h \longrightarrow 0$ and let $g(h)$ be $o(h)$ as $h \longrightarrow 0$. Show that $f(h)g(h)$ is $o(h)$ as $h \longrightarrow 0$.

2. Let $f_1(h)$ be $O(g_1(h))$ as $h \longrightarrow 0$ and let $f_2(h)$ be $o(g_2(h))$ as $h \longrightarrow 0$. Show that $f_1(h)f_2(h)$ is $o(g_1(h)g_2(h))$ as $h \longrightarrow 0$.

3. Let $f(x)$ be a function on an interval (a,b). Let $c \in (a,b)$ and let f be differentiable at c. Suppose moreover that $f'(c) > 1$. Show that there is $\delta > 0$ so that when $x \in (c,c + \delta)$, we have $f(x) > f(c)+x-c$.

4. Let f and g be functions which are n times differentiable at a point x. Denote by $f^{(j)}$ and $g^{(j)}$ the jth derivative of f and g respectively. Show that the product function fg is n times differentiable at x with

$$(fg)^{(n)}(x) = \sum_{j=0}^{n} \binom{n}{j} f^{(j)}(x)g^{(n-j)}(x).$$

Hint: Use induction.

5. Let $f(x)$ be $O(1)$ as $x \longrightarrow 0$. Show that $x^2 f(x)$ is differentiable at $x = 0$.

6. Show that $f(x) = x^{\frac{1}{3}}$ is not differentiable at $x = 0$.

Flipped recitation, Derivatives
In this section, we introduced O and o notation which allowed us to easily prove the rules of differentiation.

Problem 1
Let a and b be positive numbers. Show that x^a is $o(x^b)$ as $x \longrightarrow 0$ if and only if $a > b$.

Problem 2
Let a and b be positive numbers. Show that x^a is $O(x^b)$ as $x \longrightarrow 0$ if and only if $a \geq b$.

Problem 3
Let $p(x)$ be a polynomial with a root at r; that is, $p(r) = 0$. Show that if we define

$$q(x) = p(x - r),$$

then $q(x)$ is $O(x)$. Under what conditions is it the case that $q(x)$ is $o(x)$?

◇ **3.3   Mean Value Theorem**

In this section, we'll state and prove the mean value theorem and describe ways in which the derivative of a function gives us global information about its behavior.

Local maximum (or minimum) Let f be a real valued function on an interval $[a,b]$. Let c be a point in the interior of $[a,b]$. That is, $c \in (a,b)$. We say that f has a local maximum (respectively, local minimum) at c if there is some $\epsilon > 0$ so that $f(c) \geq f(x)$ (respectively, $f(c) \leq f(x)$) for every $x \in (c - \epsilon, c + \epsilon)$.

Lemma 3.3.1 Let f be a real valued function on $[a,b]$, differentiable at the point c of the interior of $[a,b]$. Suppose that f has a local maximum or local minimum at c. Then

$$f'(c) = 0.$$

Proof of Lemma 3.3.1 Since f is differentiable at c, we have that

$$f(x) = f(c) + f'(c)(x - c) + o(|x - c|)$$

as $x - c \longrightarrow 0$. Suppose that $f'(c) \neq 0$. From the definition of o, we have that there is some $\delta > 0$ so that

$$|f(x) - f(c) - f'(c)(x - c)| \leq \frac{|f'(c)||x - c|}{2},$$

whenever

$$|x - c| < \delta.$$

Proof of Lemma 3.3.1 cont.

(The inequality is true since the limit of $T(x,c) = |f(x) - f(c) - f'(c)(x - c)|$ divided by $|x - c|$ is 0, so by choosing $\delta(\epsilon)$ sufficiently small, we can make $|T(x,c)| \leq \epsilon|x - c|$ for any $\epsilon > 0$.) Thus whenever $|x - c| < \delta$, the sign of $f(x) - f(c)$ is the same as the sign of $f'(c)(x - c)$. This sign changes depending on whether $x - c$ is positive or negative. But this contradicts $f(c)$ being either a local maximum or a local minimum. Thus our initial assumption was false and we have $f'(c) = 0$ as desired.

In high school calculus, Lemma 3.3.1 is often used for solving optimization problems. Suppose we have a function f which is continuous on $[a,b]$ and differentiable at every point in the interior of $[a,b]$. Then from the extreme value theorem, we know the function achieves a maximum on $[a,b]$. One possibility is that the maximum is at a or at b. If this is not the case, then the maximum must be at a point where $f'(c) = 0$. Instead, we shall use lemma 3.3.1 to prove the mean value theorem.

Lemma 3.3.2 Rolle's Theorem

Let $f(x)$ be a function which is continuous on the closed interval $[a,b]$ and differentiable on every point of the interior of $[a,b]$. Suppose that $f(a) = f(b)$. Then there is a point $c \in [a,b]$ where $f'(c) = 0$.

Proof of Rolle's theorem

By the extreme value theorem, f achieves its maximum on $[a,b]$. By applying the extreme value theorem to $-f$, we see that f also achieves its minimum on $[a,b]$. By hypothesis, if both the maximum and the minimum are achieved on the boundary, then the maximum and the minimum are the same and thus the function is constant. A constant function has zero derivative everywhere. If f is not constant, then f has either a local minimum or a local maximum in the interior. By Lemma 3.3.1, the derivative at the local maximum or local minimum must be zero.

It is sometimes important to keep track of the hypotheses in our theorems. The hypotheses for Rolle's theorem and then also the mean value theorem require that the function be continuous on the closed interval but require that it be differentiable only on the interior. The reason for the continuity hypothesis is that we need to use the extreme value theorem, Theorem 3.1.2. The reason for the differentiability hypothesis is that we are only using differentiability to study local maxima on the interior.

Theorem 3.3.1 Mean value theorem

Let $f(x)$ be a function which is continuous on the closed interval $[a,b]$ and which is differentiable at every point of (a,b). Then there is a point $c \in (a,b)$ so that

$$f'(c) = \frac{f(b) - f(a)}{b - a}.$$

Proof of mean value theorem

Replace $f(x)$ by

$$g(x) = f(x) - \frac{(f(b) - f(a))(x - a)}{b - a}.$$

Observe that $g(a) = f(a)$ and $g(b) = f(b) - (f(b) - f(a)) = f(a)$. Further, g has the same continuity and differentiability properties as f since

$$g'(x) = f'(x) - \frac{f(b) - f(a)}{b - a}.$$

Thus we may apply Rolle's theorem to g to find $c \in (a,b)$, where $g'(c) = 0$. We immediately conclude that

$$f'(c) = \frac{f(b) - f(a)}{b - a},$$

proving the theorem.

We can use the mean value theorem to establish some standard ideas about the meaning of the derivative as well as some standard tests for determining whether a critical point, a point c in the interior of the domain of a function f, where $f'(c) = 0$, is a local maximum or a local minimum.

Proposition 3.3.1

Suppose a function f is continuous on the interval $[a,b]$ and differentiable at every point of the interior (a,b). Suppose that $f'(x) > 0$ for every $x \in (a,b)$. Then $f(x)$ is strictly increasing on $[a,b]$. (That is, for every $x,y \in [a,b]$ if $x < y$ then $f(x) < f(y)$.)

Proof of Proposition 3.3.1

Given $x, y \in [a,b]$ with $x < y$, we have that f satisfies the hypotheses of the mean value theorem on $[x,y]$. Thus there is $c \in (x,y)$ so that

$$f(y) - f(x) = f'(c)(y - x).$$

Since we know that $f'(c) > 0$, we conclude that

$$f(y) - f(x) > 0$$

or, in other words,

$$f(y) > f(x).$$

Thus f is increasing.

Theorem 3.3.2 First derivative test

Let f be a function continuous on $[a,b]$ and differentiable on (a,b). Let c be a point of (a,b) where $f'(c) = 0$. Suppose there is some $\delta > 0$ so that for every $x \in (c - \delta, c)$, we have that $f'(x) > 0$, and for every $x \in (c, c + \delta)$, we have that $f'(x) < 0$. Then f has a local maximum at c.

Proof of First Derivative Test

By choosing δ sufficiently small, we arrange that $(c - \delta, c + \delta) \subset (a,b)$. Thus, we may apply the proposition 3.3.1 to f on $[c - \delta, c]$ concluding that $f(c) > f(x)$ for any $x \in (c - \delta, c]$. Next, we apply proposition 3.3.1 to $-f$ on the interval $[c, c + \delta]$, concluding that $-f(x) > -f(c)$ for any $x \in (c, c + \delta]$. Multiplying the inequality by -1, we see this is the same as $f(c) > f(x)$. Thus f achieves its maximum on $[c - \delta, c + \delta]$ at c. In other words, f has a local maximum at c.

**Theorem
3.3.3
Second
derivative
test**

Let f be a function continuous on $[a,b]$ and differentiable on (a,b). Let c be a point of (a,b) where $f'(c) = 0$. Suppose the derivative $f'(x)$ is differentiable at c and that $f''(c) < 0$. Then f has a local maximum at c.

**Proof of the
second
derivative test**

Since $f'(c) = 0$, we have that

$$f'(x) = f''(c)(x - c) + o(x - c)$$

as $x \longrightarrow c$. In particular, there is $\delta > 0$ so that

$$|f'(x) - f''(c)(x - c)| < \frac{1}{2}|f''(c)||x - c|.$$

(This is true since indeed we can choose δ to bound by $\epsilon|x - c|$ for any $\epsilon > 0$.) Thus for any $x \in (c - \delta, c)$, we have that $f'(x) > 0$, while for any $x \in (c, c + \delta)$ we have $f'(x) < 0$. Thus we may apply the first derivative test to conclude that f has a local maximum at c.

Exercises for Section 3.3

1. Show that the mean value theorem fails if we replace the hypothesis that f is continuous on $[a,b]$ by simply the hypothesis that it is continuous on (a,b).

2. Let f be a function with a continuous third derivative in an open interval containing a point c. Suppose that $f'(c) = 0$ and $f''(c) = 0$. Show that if $f'''(c) \neq 0$, then f has neither a local minimum nor a local maximum at c.

3. Let f be a function with a continuous nth derivative in an open interval containing a point c. Suppose each of the first, second, ..., and $n - 1$st derivatives of f are zero at c. Suppose the nth derivative $f^{(n)}(c) \neq 0$. Does f have a local minimum, a local maximum, or neither at c? Your answer may depend on whether n is even or odd and on the sign of the nth derivative.

4. Let $p(x)$ be a polynomial on the real line with real coefficients. Suppose that $p'(x)$ has at most $d-1$ real roots. Show that $p(x)$ has no more than d real roots.

5. We say that a function f is *convex* on the interval $[a,b]$ if for every $c,d \in [a,b]$ and for every $t \in [0,1]$, we have the inequality

$$tf(c)+(1-t)f(d) \geq f(tc+(1-t)d).$$

This means that the secant line of the graph of f between c and d is above the graph. Show that if f is convex and if f is twice differentiable on (a,b), then

$$f''(x) \geq 0$$

for every $x \in (a,b)$. Hint: It is enough to prove that the derivative f' is nondecreasing on (a,b) and then to use the definition of the derivative. To prove this, assume there are points $c,d \in (a,b)$ with $c < d$ but $f'(d) < f'(c)$ and contradict convexity. You can use the continuity of the first derivative and the mean value theorem.

6. Let f be a continuous convex function on $[a,b]$ which is not constant on any subinterval of $[a,b]$. (But f is not necessarily twice or even once differentiable.) Show that f achieves a *unique* local minimum on $[a,b]$. Hint: Assume there are two local minima and reach a contradiction.

Flipped recitation, mean value theorem
In this section, we proved the mean value theorem and showed its relationship to problems in optimization.

Problem 1 Show that the polynomial

$$p(x) = 4x^3 - 18x^2 + 24x - 9$$

has three real roots. Hint: Evaluate $p(x)$ at the roots of its derivative.

Problem 2 Find the local maxima and minima of

$$p(x) = 4x^3 - 18x^2 + 24x - 9.$$

Use the first derivative test to verify that they are local maxima and minima.

Problem 3 Find the local maxima and minima of

$$p(x) = 4x^3 - 18x^2 + 24x - 9.$$

Use the second derivative test to verify that they are local maxima and minima.

◇ 3.4 Applications of the Mean Value Theorem

In the previous section, we proved the mean value theorem:

Theorem 3.4.1 Mean Value theorem again Let f be a function which is continuous on the closed interval $[a,b]$ and which is differentiable at every point of the interior (a,b). Then there is $c \in (a,b)$ so that

$$f'(c) = \frac{f(b) - f(a)}{b - a}.$$

On first glance, this seems like not a very quantitative statement. There is a point c in the interval (a,b) where the equation holds, but we can't use the theorem to guess exactly where that point c is, and so it is hard for us to use the mean value theorem to obtain information about large-scale changes in the function f from the value of its first derivative.

But this objection is somewhat misleading. The mean value theorem is really the central result in calculus, a result which permits a number of rigorous quantitative estimates. How does that work? The trick is to apply the mean value theorem primarily on intervals where the derivative of the function f is not changing too much. As it turns out, understanding second derivatives is key to effectively applying the mean value theorem. In this section we give some examples.

In an early version of this text, the theorem which follows was mistakenly given as an exercise. It is a true theorem but not as easy to prove as it looks. The idea behind the proof that will be presented here may be the deepest idea in the book.

Theorem 3.4.2

Let f be a function on an interval $[a,b]$ and let $c \in (a,b)$ be a point in the interior. Suppose that f is twice differentiable at c. (That is, suppose that f is differentiable at every point of an open interval containing c and that the derivative f' is differentiable at c.) Then

$$f''(c) = \lim_{h \longrightarrow 0} \frac{f(c+h) + f(c-h) - 2f(c)}{h^2}.$$

It is tempting to try to prove this just by comparing the first derivative of f to the difference quotients

$$\frac{f(c+h) - f(c)}{h}$$

and

$$\frac{f(c) - f(c-h)}{h},$$

subtracting them and dividing by h. These differ from $f'(c)$ and $f'(c-h)$ by $o(1)$ respectively. If we subtract the two difference quotients and divide by h, we do get the expression

$$D(c,h) = \frac{f(c+h) + f(c-h) - 2f(c)}{h^2},$$

whose difference we take the limit of in Theorem 3.4.2. However, the differentiability of f only guarantees that this difference quotient $D(c,h)$ differs from

$$\frac{f'(c) - f'(c-h)}{h}$$

by $o(\frac{1}{h})$ (because we divided by h). That is not enough to guarantee that the limit of $D(c,h)$ is the same as the second derivative. To get that, we need a little more than the differential approximation for f, which only estimates $f(c+h) - f(c)$ to within $o(h)$. We need an estimate that is within $o(h^2)$ because of the h^2 in the denominator of the expression under the limit. We will get such an estimate by using the second derivative to get a differential approximation for the first derivative and then using the mean value theorem in quite small intervals. We proceed.

Theorem 3.4.3 Taylor approximation, order 2, weak version

Let f be a function which is continuous on an interval I having c on its interior and suppose that $f'(x)$ is defined everywhere in I. Suppose further that $f''(c)$ is defined. Then for h sufficiently small that $[c, c + h] \subset I$, we have

$$f(c + h) = f(c) + hf'(c) + \frac{h^2}{2} f''(c) + o(h^2).$$

Proof of theorem 3.4.3

We see from the differential approximation for $f'(x)$ that

$$f'(c + t) = f'(c) + tf''(c) + o(t),$$

for $t < h$. Since we restrict to $t < h$, we can replace $o(t)$ by $o(h)$. (The equation above depends on both t and h.) So we record:

$$f'(c + t) = f'(c) + tf''(c) + o(h). \quad (1)$$

Now our plan is to use the expression (1) for f' together with the mean value theorem on small intervals to obtain a good estimate for $f(c + h) - f(c)$. We should specify what these intervals are going to be. We will pick a natural number n which will be the number of equal pieces into which we divide the interval $[c, c + h]$. We define points x_j where j will run from 0 to n as follows:

$$x_j = c + \frac{jh}{n}.$$

Proof of theorem 3.4.3 cont.

We observe that we can calculate $f(c + h) - f(c)$, which we are interested in, by understanding $f(x_j) - f(x_{j-1})$ for each j from 1 to n. Precisely, we have

$$f(c + h) - f(c) = \sum_{j=1}^{n} f(x_j) - f(x_{j-1}),$$

since the sum telescopes to $f(x_n) - f(x_0) = f(c+h) - f(c)$. Now we will understand each term in the sum using the mean value theorem on $[x_{j-1}, x_j]$. There is $y_j \in (x_{j-1}, x_j)$ with

$$\frac{h}{n} f'(y_j) = f(x_j) - f(x_{j-1}).$$

Thus we rewrite the sum as

$$f(c + h) - f(c) = \sum_{j=1}^{n} \frac{h}{n} f'(y_j).$$

Now we estimate $f'(y_j)$ using the differential approximation, equation (1). We conclude

$$f'(y_j) = f'(c) + (y_j - c) f''(c) + o(h).$$

Now we use the fact that $y_j \in (x_{j-1}, x_j)$ to estimate $y_j - c = \frac{jh}{n} + O(\frac{h}{n})$. Now we combine everything.

$$f(c + h) - f(c) =$$

$$\sum_{j=1}^{n} [\frac{h}{n} f'(c) + \frac{jh^2}{n^2} f''(c) + O(\frac{h^2}{n^2}) + o(\frac{h^2}{n})].$$

Proof of theorem 3.4.3 cont.

Now we sum all the terms in the square bracket separately. The terms constant in j get multiplied by n. Just the second term depends on j (linearly!) and we recall our old notation

$$S_1(n) = \sum_{j=1}^{n} j$$

to write

$$f(c+h) - f(c) =$$

$$hf'(c) + h^2 \frac{S_1(n)}{n^2} f''(c) + O(\frac{h^2}{n}) + o(h^2).$$

Observe that this equality holds for every choice of n. We take the limit as n goes to infinity and obtain the desired result, remembering that

$$\lim_{n \to \infty} \frac{S_1(n)}{n^2} = \frac{1}{2}.$$

Let's take a deep breath. A great deal happened in that argument. It might take a moment to digest. But at least you have the power to use the second-order Taylor approximation to prove Theorem 3.4.2.

Proof of theorem 3.4.2

Plugging in the weak version of Taylor approximation to order 2, we get

$$f(c+h) + f(c-h) - 2f(c) = f''(c)h^2 + o(h^2).$$

(All the lower-order terms in h cancel.) This is just what we wanted to prove.

A bit of further reflection shows that if we have enough derivatives, we can adapt the above argument to give us an estimate for $f(c+h) - f(c)$ up to $o(h^m)$ for any m. Let's do this. We adopt the notation that where defined, $f^{(j)}$ denotes the jth derivative of f.

Theorem 3.4.4 Taylor approximation, arbitrary order, weak version

Let f be a function which is continuous on an interval I having c on its interior and suppose that $f'(x), \ldots f^{(m-2)}(x)$ are defined and continuous everywhere in I. Suppose that $f^{(m-1)}$ is defined everywhere on I and that $f^{(m)}(c)$ is defined. Then for h sufficiently small that $[c, c+h] \subset I$, we have

$$f(c + h) = f(c) + \sum_{k=1}^{m} \frac{h^k}{k!} f^{(k)}(c) + o(h^m).$$

Proof of theorem 3.4.4

We will prove this approximation by induction on m. We observe that the base case $m = 2$ is just theorem 3.4.3. Thus we need only perform the induction step. We use the induction hypothesis to get an appropriate estimate for the first derivative anywhere in the interval $[c, c+h]$.

$$f'(c + t) =$$

$$f'(c) + \sum_{k=2}^{m} \frac{f^{(k)}(c)}{(k-1)!} t^{k-1} + o(h^{m-1}). \qquad (2)$$

Now we proceed as in the proof of theorem 3.4.3. We choose n and let $x_j = c + \frac{jh}{n}$. We observe that

$$f(c + h) - f(c) = \sum_{j=1}^{n} f(x_j) - f(x_{j-1}),$$

and we use the mean value theorem to find $y_j \in (x_{j-1}, x_j)$ with

Proof of theorem 3.4.4 cont.)

$$f(x_j) - f(x_{j-1}) = f'(y_j)\frac{h}{n}.$$

Now we use equation (2) to estimate $f'(y_j)$ as before and we obtain

$$f(c+h) - f(c) = \sum_{j=1}^{n}[(\sum_{k=1}^{m}(\frac{h^k}{(k-1)!n^k}f^{(k)}(c)j^{k-1}$$

$$+O(\frac{h^m}{n^2})) + o(\frac{h^m}{n})].$$

Now summing in j, we obtain

$$f(c+h) - f(c) = \sum_{k=1}^{m}\frac{h^k}{(k-1)!}f^{(k)}(c)\frac{S_{k-1}(n)}{n^k}$$

$$+O(\frac{h^m}{n}) + o(h^m).$$

Letting n tend to infinity and using the fact that

$$\lim_{n \to \infty}\frac{S_{k-1}(n)}{n^k} = \frac{1}{k},$$

we obtain the desired result.

We make some remarks:

1. You should be tempted to ask, "Does this mean that every function which is infinitely differentiable everywhere can be given as the sum of a power series?" We have shown that if the function is n times differentiable near c, then at $c+t$ near c, we have

$$f(c+t) = T_{n,f,c}(t) + o(t^n),$$

where $T_{n,f,c}$ is the degree n Taylor approximation to f near c. It is tempting to try to take the limit as $n \to \infty$. This doesn't work because of the definition of $o(t^n)$. We know that

$$\lim_{t \to 0}\frac{f(c+t) - T_{n,f,c}(t)}{t^n} = 0,$$

but for different t, the rate at which the limit goes to 0 can differ substantially.

2. Am I not then pulling a fast one in light of 1. Didn't I say in the proofs that

$$f(c + t) = f(c) + tf'(c) + o(h),$$

when in fact, it should be $o(t)$ but with a different rate of convergence for each t? No. But it's because these estimates are all coming from the *same* limit. Suppose it weren't true that

$$f(c + t) = f(c) + tf'(c) + o(h).$$

Then there is no $\delta > 0$ so that $|h| < \delta$ implies

$$|f(c + t) - f(c) - tf'(c)| \leq \epsilon h$$

for every $t < h$. Then we could pick a sequence t_j going to 0, for which the absolute value is greater than ϵh and hence ϵt_j. This would violate the definition of the derivative at c.

3. What's going on in these proofs? How come there are all these complicated sums showing up. Calculus isn't really about sums is it? There must be some way we can make them all go away. In fact, we can by subsuming them into a definition. It's a definition that's coming up in Chapter 4: the definition of the integral. What we really did in the proof of the second order version of Taylor's approximation is to calculate:

$$f(c + h) - f(c)$$

$$= \int_c^{c+h} f'(x)dx$$

$$= \int_c^{c+h} [f'(c) + xf''(c) + o(h)]dx.$$

In the same way, the kth-order version of Taylor's approximation can be obtained by integrating the $k - $1st-order version.

4. This trick of integrating is not the usual way that textbooks prove the weak version of the Taylor approximation. Instead they use L'Hopital's rule. Here is the precise statement.

Theorem 3.4.5 L'Hopital's rule

Let $f(x)$ and $g(x)$ be functions defined and continuous on an interval $[a,b]$. Suppose they are differentiable on the open interval (a,b). Suppose that $c \in (a,b)$, $f(c) = g(c) = 0$, and the limit

$$\lim_{x \to c} \frac{f'(c)}{g'(c)}$$

exists. (This requires that $g'(x)$ be different from 0 for all points but c of an open interval containing c.) Then

$$\lim_{x \to c} \frac{f(x)}{g(x)} = \lim_{x \to c} \frac{f'(c)}{g'(c)}.$$

The proof of this is again an application of the mean value theorem or, more specifically, Rolle's theorem.

Proof of L'Hopital's rule

Let x be any point different from c. We apply Rolle's theorem to the function

$$h(t) = f(x)g(t) - g(x)f(t),$$

on the interval $[c,x]$ if $x > c$ or on $[x,c]$ if $x < c$. It is easy to check (using the fact that $f(c) = g(c) = 0$) that the function h satisfies the hypotheses of Rolle's theorem. Thus there is $d \in (c,x)$ or (x,c) depending on which makes sense, so that

$$h'(d) = 0.$$

Unwinding the algebra, we get that

$$\frac{f'(d)}{g'(d)} = \frac{f(x)}{g(x)}.$$

Proof of
L'Hopital's
rule cont.

Let
$$L = \lim_{x \to c} \frac{f'(c)}{g'(c)},$$

which by hypothesis, we know exists. Then for every $\epsilon > 0$ there is $\delta(\epsilon)$ so that $|d - c| < \delta$ implies that

$$\left| L - \frac{f'(d)}{g'(d)} \right| < \epsilon.$$

Now suppose that $|x - c| < \delta$. We've shown that there is d with $|d - c| < |x - c|$ so that

$$\frac{f'(d)}{g'(d)} = \frac{f(x)}{g(x)}.$$

Thus
$$\left| L - \frac{f(x)}{g(x)} \right| < \epsilon,$$

which was to be shown.

A typical way that L'Hopital's rule can be applied is to start with a limit
$$\lim_{x \to c} \frac{f(x)}{g(x)},$$
which we can't compute because, though f and g are continous at c, we have $f(c) = 0$ and $g(c) = 0$. We keep differentiating f and g until finally at the kth step, we are no longer in this condition and the limit is

$$\frac{f^{(k)}(c)}{g^{(k)}(c)}.$$

This is exactly how we could prove the weak version of the Taylor approximation. Moreover, we could go in the reverse direction. Given f and g, we could apply the weak Taylor approximation to order k at c and obtain the limit

$$\frac{f^{(k)}(c)}{g^{(k)}(c)}.$$

However, you should not assume that the weak Taylor approximation and L'Hopital's rule are equivalent. L'Hopital's rule is actually stronger because it applies even when it does not help compute the limit. (More precisely, it applies when f and g vanish to infinite order at c, that is, when all their derivatives vanish at c.)

Calculus students typically use L'Hopital's rule as a black box. They do not think about the proof. It is a result with a clean statement that can be used repeatedly. But I don't recommend this. It is a recipe for being able to establish results without really knowing why they are true.

Exercises for Section 3.4

1. Let f and g be functions which are twice differentiable at a point c and differentiable in some interval with c in the interior. Suppose $f(c) = f'(c) = g(c) = g'(c) = 0$. Suppose that $g''(c) \neq 0$ Prove that

$$\lim_{x \to c} \frac{f(x)}{g(x)} = \frac{f''(c)}{g''(c)}$$

using Theorem 3.4.3 .

2. Let f and g be functions which are k times differentiable at a point c and $k - 1$ times differentiable in some interval with c in the interior. Suppose $f(c) = f^{(j)}(c) = g(c) = g^{(j)}(c) = 0$, for each $0 < j < k$. Suppose that $g^{(k)}(c) \neq 0$ Prove that

$$\lim_{x \to c} \frac{f(x)}{g(x)} = \frac{f^{(k)}(c)}{g^{(k)}(c)}$$

using Theorem 3.4.4.

3. Let $p(x)$ be a polynomial. For each natural number n and each $1 \leq j \leq n$, pick a number $y_{j,n}$ in the interval $[\frac{j-1}{n}, \frac{j}{n}]$. Show that

$$\lim_{n \to \infty} \frac{1}{n} \sum_{j=1}^{n} p(y_{j,n})$$

exists and is independent of the choice of $y_{j,n}$.

4. Prove the weak version of Taylor's approximation to an arbitrary order using L'Hopital's rule. (Hint: Rewrite the conclusion as saying that a certain limit is equal to zero.)

5. Prove L'Hopital's rule at infinity. Suppose $f(x)$ and $g(x)$ are continuous and differentiable functions on all the reals with

$$\lim_{x \to \infty} f(x) = \lim_{x \to \infty} g(x) = 0.$$

(Recall this means for every $\epsilon > 0$, there is $M > 0$ so that $|x| > M$ implies $|f(x)| < \epsilon$ and $|g(x)| < \epsilon$.) Suppose

$$\lim_{x \to \infty} \frac{f'(x)}{g'(x)},$$

exists. Then show that

$$\lim_{x \to \infty} \frac{f'(x)}{g'(x)} = \lim_{x \to \infty} \frac{f(x)}{g(x)}.$$

(Hint: Apply L'Hopital's rule to the functions $F(t) = f(\frac{1}{t})$ and $G(t) = g(\frac{1}{t})$. Be careful to note that you can define F and G so that they extend continuously to $t = 0$.)

Flipped recitation, weak Taylor approximation, L'Hopital's rule

In this section, we largely bypassed l'Hopital's rule by directly proving a weak form of Taylor's approximation. Then we proved L'Hopital's rule.

Problem 1

Let $p(x) = x^2 + 5x + 3$. Fix a natural number n. For each j with $1 \leq j \leq n$, fix a real number $y_{j,n}$ with $\frac{j-1}{n} \leq y_{j,n} \leq \frac{j}{n}$. Following the proof of Theorem 3.4.3 calculate

$$\lim_{n \to \infty} \sum_{j=1}^{n} \frac{1}{n} p(y_{j,n}).$$

Problem 2

Let $p(x) = x^3 + 4x + 1$. Fix a natural number n. For each j with $1 \leq j \leq n$, fix a real number $y_{j,n}$ with $\frac{j-1}{n} \leq y_{j,n} \leq \frac{j}{n}$. Following the proof of Theorem 3.4.4 calculate

$$\lim_{n \to \infty} \sum_{j=1}^{n} \frac{1}{n} p(y_{j,n}).$$

Problem 3

Let $p(x) = x^4 + x^3 + 2x^2 + 4x + 1$. Fix a natural number n. For each j with $1 \leq j \leq n$, fix a real number $y_{j,n}$ with $\frac{j-1}{n} \leq y_{j,n} \leq \frac{j}{n}$. Following the proof of Theorem 3.4.4 calculate

$$\lim_{n \to \infty} \sum_{j=1}^{n} \frac{1}{n} p(y_{j,n}).$$

◇ 3.5 Exponentiation

This section is going to focus on exponentiation, something you may consider one of the basic operations of arithmetic. However, there is a subtle limiting process that takes place when defining exponentiation which we need to fully recognize. Taking a number to an nth power when n is a natural number is just defined recursively from multiplication.

$$x^{n+1} = x^n x.$$

Defining negative integer powers is obvious enough:

Negative powers

$$x^{-n} = \frac{1}{x^n}.$$

We define rational powers similarly to the way we defined square roots.

Rational powers

Let x be a positive real number and $\frac{p}{q}$ be a positive rational (with p and q natural numbers). Then

$$x^{\frac{p}{q}} = l.u.b.\{y : y^q < x^p\}.$$

It is worth taking a moment to prove the key law of exponents for rationals:

Lemma 3.5.1
Rational law of exponents

$$x^{\frac{p}{q}} x^{\frac{r}{s}} = x^{\frac{p}{q} + \frac{r}{s}}.$$

Proof **of**
Lemma 3.5.1

We calculate

$$x^{\frac{p}{q}} x^{\frac{r}{s}}$$

$$= l.u.b.\{y : y^q < x^p\} l.u.b.\{z : z^s < x^r\}$$

$$= l.u.b.\{y : y^{qs} < x^{ps}\} l.u.b.\{z : z^{qs} < x^{rq}\}$$

$$= l.u.b.\{yz : y^{qs} < x^{ps}, z^{qs} < x^{rq}\}$$

$$= l.u.b.\{yz : (yz)^{qs} < x^{ps+rq}\}$$

$$= x^{\frac{ps+rq}{qs}}.$$

We can think of the second equality in the proof of Lemma 3.5.1 as the process of putting the exponents under a common denominator. The rest follows naturally.

Now we are ready to define x^α for $x > 1$ and α, both positive reals.

Real powers

$$x^\alpha = l.u.b.\{x^{\frac{p}{q}} : \frac{p}{q} \in \mathbf{Q}, \frac{p}{q} < \alpha\}.$$

For α negative, and $x > 1$, we can just define x^α as $\frac{1}{x^{-\alpha}}$. For $x < 1$, we can just define x^α as $\left(\frac{1}{x}\right)^{-\alpha}$.

Let's prove the exponent law for reals.

Lemma 3.5.2
Exponent law

$$x^\alpha x^\beta = x^{\alpha+\beta}.$$

Proof of Lemma 3.5.2

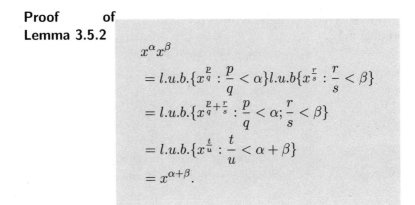

$$x^\alpha x^\beta$$

$$= l.u.b.\{x^{\frac{p}{q}} : \frac{p}{q} < \alpha\} l.u.b.\{x^{\frac{r}{s}} : \frac{r}{s} < \beta\}$$

$$= l.u.b.\{x^{\frac{p}{q}+\frac{r}{s}} : \frac{p}{q} < \alpha; \frac{r}{s} < \beta\}$$

$$= l.u.b.\{x^{\frac{t}{u}} : \frac{t}{u} < \alpha + \beta\}$$

$$= x^{\alpha+\beta}.$$

Another crucially important property of exponentiation is its continuity. To be precise:

Theorem 3.5.1

Let x be a real number $x > 1$. Consider the function $f(\alpha) = x^\alpha$. The function f is continuous at every real α.

The restriction to $x > 1$ is only for convenience. We obtain the case $x < 1$ from the limit law for quotients.

Proof of The- We need only prove that for fixed α and for
orem 3.5.1   every $\epsilon > 0$, there is a δ so that $|\beta - \alpha| < \delta$
implies that $|x^\beta - x^\alpha| < \epsilon$.
We easily calculate that

$$|x^\beta - x^\alpha| \le |x^\alpha||x^{\beta-\alpha} - 1|.$$

We claim it suffices to find δ so that $|x^\delta - 1| <$
$\frac{\epsilon}{2x^\alpha}$. Suppose we have done this. Then if $\beta - \alpha$
is positive, we immediately have

$$|x^\beta - x^\alpha| \le |x^\alpha||x^{\beta-\alpha} - 1| < \frac{\epsilon}{2}.$$

On the other hand, if $\beta - \alpha$ is negative, we
use

$$x^{\beta-\alpha} > \frac{1}{1 - \frac{\epsilon}{2x^\alpha}} > 1 - \frac{\epsilon}{x^\alpha}$$

from the geometric series as long as $\epsilon < 1$.
Thus we have

$$|x^\beta - x^\alpha| \le |x^\alpha||x^{\beta-\alpha} - 1| < \epsilon.$$

To see that such a δ exists, we simply observe
that $(1+\epsilon)^q > 1 + q\epsilon$. We pick q large enough
that $1 + q\epsilon$ is larger than x and let $\delta = \frac{1}{q}$.

Now we are ready to discuss the most important aspect
of the theory of exponentiation. That is the definition of the
natural base, the very special number e.

The most natural definition is

Definition of e

$$e = \lim_{n \to \infty} (1 + \frac{1}{n})^n.$$

One interpretation of this is that e is the number of dollars
we would have in a savings account if one year earlier we has
deposited one dollar, and the account earned interest over the
course of the year at the fantastical rate of 100%, compounded

continuously. Then the nth term of the limiting sequence is the amount of money we would have if the interest only compounded at n equal intervals.

How do we know the limit exists? We just use the binomial theorem. We calculate

$$(1 + \frac{1}{n})^n = \sum_{j=0}^{\infty} \binom{n}{j} \frac{1}{n^j}.$$

(We have no problem letting this sum run to infinity as long as we interpret $\binom{n}{j}$ to be 0 as long as $j > n$.) Now we examine the jth term in the sum. It is

$$\frac{n(n-1)\ldots(n+1-j)}{j!n^j}.$$

Clearly this increases to $\frac{1}{j!}$ as n increases to infinity. Thus we conclude that

$$e = 1 + 1 + \frac{1}{2!} + \cdots + \frac{1}{j!} + \ldots$$

This series converges easily by the ratio test.

We observe that we can rewrite the limit as the limit of a function whose argument approaches infinity rather than as the limit of a sequence.

Proposition 3.5.1

$$e = \lim_{y \to \infty} (1 + \frac{1}{y})^y.$$

We quickly sketch the proof of Proposition 3.5.1. Suppose $n < y < n + 1$. Then

$$(1 + \frac{1}{n+1})^n \leq (1 + \frac{1}{y})^y \leq (1 + \frac{1}{n})^{n+1}.$$

Then we see we have the desired limit by the squeeze theorem.

We shall be very interested in studying the function

$$f(x) = e^x.$$

What is this? By continuity of exponentiation, we get

$$e^x = [\lim_{y \to \infty} (1 + \frac{1}{y})^y]^x = \lim_{y \to \infty} (1 + \frac{1}{y})^{yx}.$$

Now we make a change of variables, introducing $z = yx$. We conclude that

$$e^x = \lim_{z \to \infty} (1 + \frac{x}{z})^z.$$

We restrict to integer z, obtaining

$$e^x = \lim_{n \to \infty} (1 + \frac{x}{n})^n.$$

We can analyze the last expression as we did the expression for e: by using the binomial theorem. The jth term converges to $\frac{1}{j!}x^j$. We conclude

$$e^x = \sum_{j=0}^{\infty} \frac{x^j}{j!}.$$

Thus we have a power series expression for the function e^x.

We can differentiate the function e^x from the definition of the derivative:

$$\frac{d}{dx} e^x$$

$$= \lim_{h \to 0} \frac{e^{x+h} - e^x}{h}$$

$$= e^x \lim_{h \to 0} \frac{e^h - 1}{h}$$

$$= e^x \lim_{h \to 0} \frac{(\sum_{j=0}^{\infty} \frac{h^j}{j!}) - 1}{h}$$

$$= e^x \lim_{h \to 0} \sum_{j=1}^{\infty} \frac{h^{j-1}}{j!}$$

$$= e^x.$$

In the next section, we will use the remarkable properties of the function e^x to obtain negative results on the question of when infinitely differentiable functions are given by convergent Taylor series.

Exercises for Section 3.5

1. Prove for α, an irrational real number, that when $f(x) = x^\alpha$ and when $x > 0$, we have that

$$f'(x) = \alpha x^{\alpha-1}.$$

Hint: Use the definition of the derivative as a limit and be prepared to use the definition of the limit. Compare x^α to x^r with r rational. Use different values of r for different values of h in the definition of the derivative. You may have to use the mean value theorem and the continuity of $x^{\alpha-1}$. Shorter proofs are perhaps possible.

2. Prove that

$$e^x e^y = e^{x+y},$$

by writing each of e^x and e^y as its power series and multiplying them.

3. Show that $(1+\frac{1}{j})^j < e$ for any j. Conclude that

$$(\frac{j+1}{j})^n \le e^{\frac{n}{j}}.$$

4. Prove that

$$\lim_{x \longrightarrow \infty} x^n e^{-x} = 0$$

without using L'Hopital's rule. Hint: e^m is the product of the constant e, m times. Write m^n as a product of m factors of the form $(\frac{j}{j-1})^n$. Show that for j large enough this is smaller than e. Use the previous exercise.

5. Give a complete proof of Proposition 3.5.1 .

Flipped recitation, exponentiation
In this section, we discussed exponentiation and the exponential function.

Problem 1
In general, is this true?

$$a^{(b^c)} = (a^b)^c.$$

Problem 2
Show that

$$\lim_{n \longrightarrow \infty} \frac{e^{(n^{1000})}}{e^{(e^n)}} = 0.$$

Problem 3
Show that $e^{\left(\frac{-1}{x^2}\right)}$ is $o(x^{1000} e^{\left(\frac{-1}{x}\right)})$ as x goes to 0.

◇ 3.6   Smoothness and series

In section 3.4, we built up the notion of weak Taylor approximation culminating with theorem 3.4.4 which we restate here.

Theorem 3.6.1
Weak Taylor approximation again.
Let $f, f', f'', \ldots, f^{(n-2)}$ be defined and continuous everywhere on a closed interval I having c in the interior. Let $f^{(n-1)}$ be defined everywhere on I and let $f^{(n)}(c)$ be defined. Then

$$f(x) = f(c) + \sum_{k=1}^{n} \frac{f^{(k)}(c)(x-c)^k}{k!} + o(|x-c|^n).$$

Suppose for a moment that a function and all (infinitely many) of its derivatives are defined and continuous on a closed interval I containing c. We say that a function f with this property is in the class $C^\infty(I)$. It is infinitely continuously differentiable on I. Then we define the power series (in $x - c$)

$$f(c) + \sum_{k=1}^{\infty} \frac{f^{(k)}(c)(x-c)^k}{k!}$$

to be the *formal Taylor series* of f at c.

We have already met some important functions given by convergent power series (which in light of theorem 3.4.4 can be shown to be their formal Taylor series), to wit,

$$\frac{1}{1-x} = 1 + x + x^2 + \cdots + x^n + \ldots \quad |x| < 1$$

We know the above because it is the formula for the sum of a geometric series.

Further,

$$e^x = 1 + x + \frac{x^2}{2} + \cdots + \frac{x^n}{n!} + \ldots,$$

which we discovered in the previous section using the binomial theorem. Moreover, the failure of the first series to converge outside of radius 1 can be explained by the fact that the function $\frac{1}{1-x}$ really has a discontinuity at $x = 1$. We could be tempted to expect that any function in $C^\infty(I)$ can be given by a convergent power series at least in small parts of that interval. In the present section, we will see why such an expectation would be wrong.

We're going to use properties of the exponential function to see why not all C^∞ functions can be given as convergent power series. An important feature of exponential growth is that it is fast. It is indeed faster than any polynomial. To be precise:

Lemma 3.6.1

$$\lim_{x \longrightarrow \infty} \frac{x^k}{e^x} = 0$$

for all natural numbers k.

The lemma is usually proven using L'Hopital's rule. But the fact is much more visceral and basic.

Proof **of** **Lemma 3.6.1** We first study this limit for x restricted to natural numbers. That is, we study

$$\frac{n^k}{e^n}.$$

Now the denominator is rather easily understood as a product of n copies of e.

$$e^n = ee\ldots e.$$

To properly compare it to n^k, we should express n^k as a product of n terms. We can readily do this by setting up a telescoping product.

$$n^k = 1(\frac{2^k}{1^k})(\frac{3^k}{2^k})\ldots(\frac{n^k}{(n-1)^k}).$$

Proof   **of** Now it is rather easy to understand the limit
Lemma **3.6.1** of the factors of n^k, that is, to show
cont.

$$\lim_{n \to \infty} \frac{n^k}{(n-1)^k} = 1.$$

Using the definition of the limit, we see that
there is $N > 0$ so that when $n > N$, we have

$$\frac{n^k}{(n-1)^k} \leq \frac{e}{2}.$$

Hence we get

$$0 \leq \frac{n^k}{e^n} = \frac{1}{e} \frac{\frac{2^k}{1^k}}{e} \cdots \frac{\frac{n^k}{(n-1)^k}}{e} \leq C(\frac{1}{2})^{n-N}.$$

Here C is the product of the first N factors
and we have just used that the rest is less than
$\frac{1}{2}$.
Now we just apply the squeeze theorem. To
control non-integer x, we assume $n < x <
n + 1$ and see

$$\frac{n^k}{e^{n+1}} \leq \frac{x^k}{e^x} \leq \frac{(n+1)^k}{e^n}.$$

We apply the squeeze theorem again.

So what did any of that have to do with the failure of formal
Taylor series to converge to their function?
Here is a very special function that we can use:

$$f(x) = \begin{cases} e^{-\frac{1}{x^2}} & \text{for } x > 0 \\ 0 & \text{for } x \leq 0 \end{cases}$$

This f is a piecewise defined function, and so we can expect
there to be problems either with the existence or the continuity
of the derivatives at $x = 0$. But let's see what happens.

$$\frac{d}{dx}(e^{-\frac{1}{x^2}}) = (\frac{2}{x^3})e^{-\frac{1}{x^2}}.$$

$$\frac{d^2}{dx^2}(e^{-\frac{1}{x^2}}) = [\frac{d}{dx}(\frac{2}{x^3})]e^{-\frac{1}{x^2}} + (\frac{2}{x^3})^2 e^{-\frac{1}{x^2}}.$$

And so on. But in general

$$\frac{d^n}{dx^n}(e^{-\frac{1}{x^2}}) = p_n(\frac{1}{x})e^{-\frac{1}{x^2}},$$

with p_n a polynomial. We use the change of variables $y = \frac{1}{x}$ to calculate that

$$\lim_{x \to 0} p_n(\frac{1}{x})e^{-\frac{1}{x^2}} = \lim_{y \to \infty} \frac{p_n(y)}{e^{y^2}} = 0.$$

But what about right at 0? Are the derivatives of f defined?

Proposition 3.6.1

$$e^{-\frac{1}{x^2}} \text{ is } o(x).$$

Proof of Proposition 3.6.1

$$\lim_{h \to 0} \frac{e^{-\frac{1}{h^2}}}{h} = \lim_{y \to \infty} \frac{y}{e^{y^2}} = 0.$$

Hence $f'(0) = 0$ by proposition 3.6.1 and the definition of the derivative. We have shown that f has a continuous derivative everywhere. Proceeding by induction to prove that $f^{(n)}(x)$ is continuous, we observe that $f^{(n-1)}(x)$ is $o(x)$ in the same way, obtaining $f^{(n)}(x) = 0$.

Putting together all we have learned about f, we obtain the following result:

Theorem 3.6.2

Our special function f is in the class $C^\infty(\mathbf{R})$, and all the derivatives $f^{(n)}(0)$ are equal to 0. Thus the formal Taylor series of f at 0 is identically 0.

Thus taking the formal power series of f at 0 throws away all information about f.

Once a mathematical hope fails, it tends to fail catastrophically. Having discovered this single weird function f, we can use it to engineer a whole menagerie of weird functions, which we proceed to do.

Theorem 3.6.3   For any closed interval $[a,b]$, there is a function $f_{[a,b]}$ which is of the class $C^\infty(\mathbf{R})$ so that $f_{[a,b]}(x) > 0$ for $x \in (a,b)$ but $f(x) = 0$ otherwise.

Proof of Theorem 3.6.3   Define

$$f_{[a,b]}(x) = f(x-a)f(b-x).$$

Corollary 3.6.1   For every interval $[-\epsilon,\epsilon]$, there is f_ϵ so that $f_\epsilon(x) = 1$ if $x \in (-\epsilon,\epsilon)$, so that $0 < f(x) \leq 1$ if $x \in (-2\epsilon,2\epsilon)$ and $f(x) = 0$ otherwise.

Proof of Corollary 3.6.1   We define

$$f_\epsilon(x) = \frac{f_{[-2\epsilon,2\epsilon]}(x)}{f_{[-2\epsilon,2\epsilon]}(x) + f_{[-3\epsilon,-\epsilon]}(x) + f_{[\epsilon,3\epsilon]}(x)}.$$

We will conclude with an interesting relationship between $C^\infty(\mathbf{R})$ functions and formal power series. We recall that there are plenty of power series whose radius of convergence is 0.

Example 1   Here's a power series with radius of convergence 0.

Consider the power series

$$\sum_{n=0}^{\infty} n!x^n.$$

It has radius of convergence 0. We can see this since the ratios of consecutive terms are nx. For any fixed $x \neq 0$, these tend to infinity.

We now state and sketch the proof of a theorem of Borel which says that nonetheless, every power series is the formal Taylor series of some $C^\infty(\mathbf{R})$ function.

Theorem 3.6.4 Borel's theorem

Let $\sum_{n=0}^\infty a_n x^n$ be some power series (any power series whatsoever). There is a $C^\infty(\mathbf{R})$ function g which has this series as its Taylor series at 0.

Sketch of Proof of Borel's theorem

Pick $\{\epsilon_k\}$, a fast-decreasing sequence of positive real numbers. (How fast will depend on the sequence $\{a_n\}$. Deciding how fast is most of the difficulty in writing a fully rigorous proof). Define

$$g(x) = \sum_{k=0}^\infty a_k x^k f_{\epsilon_k}(x).$$

Thus the kth term of the series really is exactly $a_k x^k$ for x sufficiently small. (But how small will depend on k.) On the other hand, for each non-zero x, only finitely many terms are non-zero, so the sum converges there. We simply choose the sequence $\{\epsilon_k\}$ small enough that

$$g(x) - \sum_{k=0}^n a_k x^k = o(x^n)$$

for every n.

Exercises for Section 3.6

1. Show that the function $f(x)$ which is equal to 0 when $x = 0$ and equal to $x^2 \sin(\frac{1}{x})$ otherwise is differentiable at $x = 0$ but the derivative is not continuous. Hint: Probably you're upset that we haven't defined the sine function yet, but that isn't the point. You may assume that the sine function and cosine function are continuous, that they are bounded and that the cosine function is the derivative of the sine function.

2. Find a two-sided sequence of C^∞ functions ψ_j (that is, the index j runs over all integers) on $(0,\infty)$, so that each function ψ_j is zero outside an interval $[a_j, b_j]$ with a_j and b_j positive real numbers, so that for each j, $\psi_j(x) = \psi_{j+1}(2x)$ and so that

$$\sum_{j=-\infty}^{\infty} \psi_j(x) = 1$$

for every $x \in (0,\infty)$.

3. Find a C^∞ function on the real line whose power series at 0 is the power series in example 1 of this section.

4. Give a proof of Borel's theorem in full detail. Hint: Explain exactly how to pick the numbers ϵ_k in terms of the a_k so that you get

$$g(x) - \sum_{k=0}^{n} a_k x^k = o(x^n).$$

First consider what to do when $|a_k|$ is an increasing sequence.

Flipped recitation, bump functions

In this section, we learned about some nice functions which are infinitely differentiable but vanish outside prescribed domains.

Problem 1

Find a C^∞ function on the real line which is strictly positive on $(0,1)$, identically zero on $[1,2]$ and $[-\infty,0]$, and is strictly negative on $(2,\infty)$

Problem 2

Find a C^∞ function on the real line which is equal to 1 on $[3,5]$ and is 0 on $[-\infty,1]$ and $[7,\infty]$.

Problem 3

Find a sequence of nonnegative C^∞ functions on the real line ϕ_j so that each ϕ_j is zero outside $[j-1,j+2]$ and so that

$$\sum_{j=-\infty}^{\infty} \phi_j(x) = 1$$

for every real x. Hint: If you can't get the last condition satisfied, divide by the sum.

◇ **3.7   Inverse function theorem**

In this section, we'll begin with a classical application of calculus: obtaining numerical solutions to equations.

Situation: We would like to solve the equation

$$f(x) = 0.$$

Here f is a function and we should imagine that its first and second derivatives are continuous in some interval. We approach this problem by what is usually called Newton's method.

First we make an initial guess: x_0. Probably we are not too lucky and

$$f(x_0) \neq 0.$$

Then what we do is that we calculate $f'(x_0)$. We obtain the linear approximation

$$f(x) \approx f(x_0) + f'(x_0)(x - x_0).$$

We solve for the x_1 which makes the linear approximation equal to zero. That is,

$$x_1 = x_0 - \frac{f(x_0)}{f'(x_0)}.$$

Usually, we are still not lucky and

$$f(x_1) \neq 0.$$

We obtain x_2 from x_1 in the same way, and so on. For general j, we get

$$x_j = x_{j-1} - \frac{f(x_{j-1})}{f'(x_{j-1})}.$$

A question we should ask, which is extremely practical, is how fast the sequence $\{f(x_j)\}$ converges to 0. If we knew that, we should really know how many steps of Newton's method we should have to apply to get a good approximation to a zero.

This is a job for the mean value theorem. In the interval $[x_0, x_1]$ (or $[x_1, x_0]$, depending on which of x_0 or x_1 is greater) there is a point c so that

$$f'(c) = \frac{f(x_1) - f(x_0)}{x_1 - x_0}.$$

Thus
$$f(x_1) - f(x_0) = -\frac{f(x_0)}{f'(x_0)} f'(c).$$

Suppose that in our interval where all the action is taking place we have an upper bound M for $|f''(x)|$. Then

$$
\begin{aligned}
|f(x_1)| &\le |f(x_0)| |1 - \frac{f'(c)}{f'(x_0)}| \\
&= \frac{|f(x_0)|}{|f'(x_0)|} |f'(x_0) - f'(c)| \\
&\le \frac{M|f(x_0)|}{|f'(x_0)|} |x_0 - c| \\
&\le \frac{M|f(x_0)|^2}{|f'(x_0)|^2}.
\end{aligned}
$$

Suppose further that in our interval we have a lower bound on the absolute value of the derivative,

$$|f'(x)| > \frac{1}{K}.$$

Then we conclude that

$$|f(x_1)| \le K^2 M |f(x_0)|^2.$$

Moreover, we have the same thing for every j.

$$|f(x_j)| \le K^2 M |f(x_{j-1})|^2.$$

To get any benefit from these inequalities, we must have $|f(x_0)| < \frac{1}{K^2 M}$. In other words, our initial guess should be pretty good. But, if this is true, and $|f(x_0)| = \frac{r}{K^2 M}$ with $r < 1$, we get from these inequalities:

$$|f(x_1)| \le \frac{r^2}{K^2 M},$$

$$|f(x_2)| \le \frac{r^4}{K^2 M},$$

and, in general,

$$|f(x_j) \le \frac{r^{2^j}}{K^2 M}.$$

This is a pretty fast rate of convergence. It is double exponential.

We encapsulate all of this as a theorem:

Theorem 3.7.1
Newton's method
rate of convergence

Let I be an interval and f a function which is twice continuously differentiable on I. Suppose that for every $x \in I$, we have $|f''(x)| < M$ and $|f'(x)| > \frac{1}{K}$. Then if we pick $x_0 \in I$, and we define the sequence $\{x_j\}$ by

$$x_j = x_{j-1} - \frac{f(x_{j-1})}{f'(x_{j-1})},$$

and if we assume further that each x_j is in I and that $|f(x_0)| < \frac{r}{K^2 M}$, then we obtain the estimate

$$|f(x_j)| \leq \frac{r^{2^j}}{K^2 M}.$$

Just to know an equation has a solution, we often need a lot less. We say that a function f is *strictly increasing* on an interval $[a,b]$ if for every $x,y \in [a,b]$ with $x < y$, we have $f(x) < f(y)$.

Theorem 3.7.2

Let f be continuous and increasing on $[a,b]$. Then f has an inverse uniquely defined from $[f(a), f(b)]$ to $[a,b]$.

Proof of Theorem 3.7.2

For $c \in [f(a), f(b)]$, we want x with $f(x) = c$. Since

$$f(a) \leq c \leq f(b),$$

and f is continuous, we have that there exists such an x by the intermediate value theorem. Because f is strictly increasing, this c is unique.

With f as above, if f is differentiable at a point x with a nonzero derivative, we will show that its inverse is differentiable at $f(x)$.

Theorem 3.7.3 Inverse function theorem

Let f be a strictly increasing continuous function on $[a,b]$. Let g be its inverse. Suppose $f'(x)$ is defined and nonzero for some $x \in (a,b)$. Then g is differentiable at $f(x)$ and

$$g'(f(x)) = \frac{1}{f'(x)}.$$

Proof of inverse function theorem

By the differentiability of f at x, we get

$$f(y) = f(x) + f'(x)(y - x) + o(y - x).$$

Now we solve for $y - x$.

$$(y - x) = \frac{f(y) - f(x)}{f'(x)} + o(y - x).$$

We rewrite this as

$$g(f(y)) - g(f(x)) = \frac{f(y) - f(x)}{f'(x)} + o(y - x).$$

Finally we simply observe that anything that is $o(y - x)$ is also $o(f(y) - f(x))$ since

$$\lim_{y \to x} \frac{f(y) - f(x)}{y - x} = f'(x).$$

Thus we have obtained our desired result.

Example 1   Let's invert the exponential function.
Behold the log The function e^x is strictly increasing on the

whole real line. Thus it has an inverse from the positive reals to the real line. We call this inverse function log. We have

$$\frac{d}{dx}(\log x) = \frac{1}{x}.$$

Example 2   Application: logarithmic differentiation

An important fact about log is that

$$\log(ab) = \log a + \log b.$$

This gives rise to a nice way of thinking of the product and quotient rules (and generalizations).

Instead of calculating $\frac{d}{dx}(fg)$, we calculate $\frac{d}{dx}(\log fg)$.
We get

$$\frac{d}{dx}(\log fg) = \frac{\frac{d}{dx}(fg)}{fg},$$

but on the other hand,

$$\frac{d}{dx}(\log fg) = \frac{d}{dx}(\log f + \log g) = \frac{f'}{f} + \frac{g'}{g}.$$

Solving, we get

$$\frac{d}{dx}(fg) = (\frac{f'}{f} + \frac{g'}{g})fg = f'g + g'f.$$

The same idea works for arbitrarily long products and quotients.

Exercises for Section 3.7

1. Let $\epsilon > 0$. Show that

$$\lim_{x \to 0} |x|^\epsilon \log|x| = 0.$$

2. Let f be a twice continuously differentiable function on the real line with $f'(x) > 1$ for every value of x and $|f''(x)| \le 1$ for every value of x. Let x_0 be a real number with $f(x_0) < \frac{1}{2}$ and for each natural number j let x_j be the result of the jth step of Newton's method. What is the smallest value of j for which you can guarantee that

$$|f(x_j)| \le 10^{-100}.$$

3. Let $\log_a x$ be the inverse function of a^x. Prove that

$$\log_a x = \frac{\log x}{\log a}.$$

4. Let $f(x)$, $g(x)$, $h(x)$, and $k(x)$ be non-vanishing differentiable functions. Use logarithmic differentiation to compute the derivative of

$$F(x) = \frac{f(x)g(x)h(x)}{k(x)}.$$

Flipped recitation, Newton's method

In this section we developed Newton's method for solving equations numerically and applied it to the theory of inverse functions. You may use a calculator, as in each problem you will apply Newton's method to a given function with a given starting point and a given number of steps, looking for a zero.

Problem 1 (5 steps. $x_0 = 2$)

$$f(x) = x^3 - 4x + 2$$

Problem 2 (10 steps. $x_0 = 5$)

$$f(x) = e^x - 1.$$

Problem 3 (8 steps. $x_0 = 4$)

$$f(x) = \frac{x - 3}{x + 5}.$$

Chapter 4

INTEGRATION

◇ 4.1 Definition of the Riemann integral

Our goal for this section is to begin work on integration. In particular, we would like to define $\int_a^b f(x)dx$, the definite Riemann integral of a function f on the interval $[a,b]$. Here f should be, at least, defined and bounded on $[a,b]$.

Informally, the meaning we would like to assign to $\int_a^b f(x)dx$ is area under the curve $y = f(x)$ between the vertical lines $x = a$ and $x = b$. But we'll have to come to terms with understanding what that means and gain at least some idea about which curves have a well-defined area under them.

Classically, we understand what the area of a rectangle is and not much else. (Parallelograms are rectangles with the same triangle added and subtracted. Triangles are half-parallelograms. The areas of all these objects are built up from the area of a rectangle.) Our idea will be that we will study certain unions of disjoint rectangles contained in the region under the curve, whose areas we will call *lower Riemann sums*, and we will study unions of disjoint rectangles covering the region, whose areas we will call *upper Riemann sums*, and our integral will be defined when we can squeeze the area of the region tightly between the upper and lower sums. **Warning:** Being able to do this will put some restrictions on f.

Partition of an interval   Given an interval $[a,b]$, a partition $\mathcal{P}$ of $[a,b]$ is a set of points $\{x_0, \ldots, x_n\}$ so that

$$x_0 = a < x_1 < x_2 \cdots < x_{n-1} < x_n = b.$$

144

A partition of an interval is a set of points which act as dividers of the interval. We often think of this as the partition dividing the interval into subintervals.

Subintervals of a partition With $\mathcal{P}$ as above, the subintervals of $\mathcal{P}$ are the intervals $[x_{j-1}, x_j]$ where j runs from 1 to n.

Refinement of a partition We say a partition $\mathcal{Q}$ refines the partition $\mathcal{P}$ provided that

$$\mathcal{P} \subset \mathcal{Q},$$

that is, provided every point of $\mathcal{P}$ is also a point of $\mathcal{Q}$.

A partition $\mathcal{Q}$ refines a partition $\mathcal{P}$ if every subinterval of $\mathcal{Q}$ is contained in a subinterval of $\mathcal{P}$.

Riemann upper sum Given a partition $\mathcal{P}$ of $[a,b]$ and f a bounded function on $[a,b]$, we define

$$U_{\mathcal{P}}(f) =$$

$$\sum_{j=1}^{n} l.u.b.\{f(x) : x_{j-1} \leq x \leq x_j\}(x_j - x_{j-1}),$$

the *Riemann upper sum* of f with respect to the partition $\mathcal{P}$.

Here $l.u.b.$ denotes least upper bound.

Given a set A of real numbers bounded below, we define its $g.l.b.$ (*greatest lower bound*) by

$$g.l.b.(A) = -l.u.b.(-A),$$

where $-A$ is the set of negatives of elements of A. If A is bounded below, then $g.l.b(A)$ is defined because the negative of a lower bound for A is an upper bound for $-A$.

**Riemann
lower sum**

Now, we define

$$L_{\mathcal{P}}(f) =$$

$$\sum_{j=1}^{n} g.l.b.\{f(x) : x_{j-1} \le x \le x_j\}(x_j - x_{j-1}),$$

the *Riemann lower sum* of f with respect to the partition $\mathcal{P}$.

We record some facts about Riemann upper and lower sums.

**Proposition
4.1.1**

Let $\mathcal{P} = \{x_0, x_1, \ldots, x_n\}$ be a partition of $[a,b]$ and let $\mathcal{Q}$ be a partition which refines $\mathcal{P}$. Then for any bounded f defined on $[a,b]$, we have

$$L_{\mathcal{P}}(f) \le L_{\mathcal{Q}}(f) \le U_{\mathcal{Q}}(f) \le U_{\mathcal{P}}(f).$$

**Proof   of
Proposition
4.1.1**

We observe that for every pair of adjacent points of $\mathcal{P}$ namely x_{j-1}, x_j, the subset $\mathcal{Q}_{[x_{j-1},x_j]}$ consisting of points in $\mathcal{Q}$ contained in $[x_{j-1},x_j]$ is a partition of $[x_{j-1},x_j]$. It suffices to show that

$$g.l.b.\{f(x) : x_{j-1} \le x \le x_j\}(x_j - x_{j-1}) \le$$

$$L_{\mathcal{Q}_{[x_{j-1},x_j]}} \le U_{\mathcal{Q}_{[x_{j-1},x_j]}}$$

$$\le l.u.b.\{f(x) : x_{j-1} \le x \le x_j\}(x_j - x_{j-1}).$$

This is true because the g.l.b.'s in the definition of $L_{\mathcal{Q}_{[x_{j-1},x_j]}}$ are all larger than or equal to the g.l.b. on all of $[x_{j-1},x_j]$, which in turn are smaller than or equal to the respective l.u.b.'s, which are smaller than or equal to the l.u.b. on all of $[x_{j-1},x_j]$. Now we just sum our inequalities over j to obtain the desired result.

Corollary 4.1.1 Let $\mathcal{P}$ and $\mathcal{Q}$ be any partitions of $[a,b]$. Then for any bounded f on $[a,b]$,

$$L_{\mathcal{P}}(f) \leq U_{\mathcal{Q}}(f).$$

Proof of Corollary 4.1.1 Clearly $\mathcal{P} \cup \mathcal{Q}$ refines both $\mathcal{P}$ and $\mathcal{Q}$. We simply use Proposition 4.1.1 to show that

$$L_{\mathcal{P}}(f) \leq L_{\mathcal{P} \cup \mathcal{Q}}(f) \leq U_{\mathcal{Q}}(f).$$

Lower integral Thus we have obtained that the set of all lower Riemann sums of a bounded function on $[a,b]$ are bounded above, and we denote

$$l.u.b.\{L_{\mathcal{P}}(f)\} = I_{l,[a,b]}(f),$$

where the l.u.b. is taken over all partitions of $[a,b]$. We call $I_{l,[a,b]}(f)$ the *lower integral* of f on $[a,b]$.

Upper integral Similarly the upper sums are all bounded below. We denote

$$g.l.b.\{U_{\mathcal{P}}(f)\} = I_{u,[a,b]}(f),$$

where the g.l.b. is taken over all partitions of $[a,b]$. We call $I_{u,[a,b]}(f)$ the *upper integral* of f on $[a,b]$.

Riemann integral

When the two numbers $I_{l,[a,b]}(f)$ and $I_{u,[a,b]}(f)$ are equal, we say that f is Riemann integrable on $[a,b]$ and we call this common number

$$\int_a^b f(x)dx.$$

At last we have completed the definition of the integral.

Example 1   **Warning: Not every bounded function is Riemann integrable.**

Proof

Let $f(x)$ be defined on $[0,1]$ by

$$f(x) = \begin{cases} 1 & \text{for } x \in \mathbf{Q} \\ 0 & \text{for } x \notin \mathbf{Q}. \end{cases}$$

Here $\mathbf{Q}$ denotes the rational numbers.

It is easy to see that any upper sum of f on $[0,1]$ is 1 (since every subinterval contains a rational number) and any lower sum is 0 (since every subinterval contains an irrational number.) The function f is not Riemann integrable. There are more sophisticated integrals that can handle this f but no system of integration will work on any function.

We record some basic properties of Riemann integration:

Theorem 4.1.1

Let f, g be Riemann integrable on $[a,b]$ and c_1, c_2, c, and $k \neq 0$ be numbers.

(i) $\displaystyle \int_a^b c_1 f + c_2 g = c_1 \int_a^b f + c_2 \int_a^b g.$

(ii) $\displaystyle \int_a^b f(x)dx = \int_{a+c}^{b+c} f(x-c)dx.$

(iii) $\displaystyle \int_a^b f(x)dx = \frac{1}{k} \int_{ka}^{kb} f\left(\frac{x}{k}\right)dx.$

If for every $x \in [a,b]$, we have $g(x) \leq f(x)$, then

(iv) $\displaystyle \int_a^b g(x)dx \leq \int_a^b f(x)dx.$

If $c \in [a,b]$,

(v) $\displaystyle \int_a^c f(x)dx + \int_c^b f(x)dx = \int_a^b f(x)dx.$

We proceed to prove the parts of Theorem 4.1.1.

Proof of Theorem 4.1.1 (i)

We let $\mathcal{P}$ be any partition of $[a,b]$. We restrict to the case that c_1, c_2 are nonnegative. We see that

$$U_{\mathcal{P}}(c_1 f + c_2 g) \leq c_1 U_{\mathcal{P}}(f) + c_2 U_{\mathcal{P}}(g),$$

since the maximum (or near maximum) in every subinterval of the partition may occur at different points for f and g. Similarly,

$$c_1 L_{\mathcal{P}}(f) + c_2 L_{\mathcal{P}} g \leq L_{\mathcal{P}}(c_1 f + c_2 g).$$

Taking respectively the l.u.b and the g.l.b., we get

$$I_{u,[a,b]}(c_1 f + c_2 g) \leq c_1 \int_a^b f + c_2 \int_a^b g$$

and

$$I_{l,[a,b]}(c_1 f + c_2 g) \geq c_1 \int_a^b f + c_2 \int_a^b g.$$

Since

$$I_{l,[a,b]}(c_1 f + c_2 g) \leq I_{u,[a,b]}(c_1 f + c_2 g),$$

we have shown that $c_1 f + c_2 g$ is Riemann integrable and that (i) holds. To get the full power of (i), we must consider negative c_1 and c_2. It is enough to show that if f is integrable on $[a,b]$, then so is $-f$. We see immediately that

$$I_{u,[a,b]}(f) = -I_{l,[a,b]}(-f)$$

and

$$I_{l,[a,b]}(f) = -I_{u,[a,b]}(-f).$$

Thus $-f$ is integrable with integral $-\int_a^b f$.

Proof of Theorem 4.1.1 (ii)

We see that any partition $\mathcal{P}$ of $[a,b]$ can be transformed into a partition $\mathcal{P}+c$ of $[a+c,b+c]$ (and *vice versa*) and we see that

$$U_{\mathcal{P}}(f(x)) = U_{\mathcal{P}+c}(f(x-c))$$

and

$$L_{\mathcal{P}}(f(x)) = L_{\mathcal{P}+c}(f(x-c)).$$

Proof of Theorem 4.1.1 (iii)

Similarly, we see that any partition $\mathcal{P}$ of $[a,b]$ can be transformed into a partition $k\mathcal{P}$ of $[ka,kb]$ (and *vice versa*) and we see that

$$U_{\mathcal{P}}(f(x)) = \frac{1}{k}U_{k\mathcal{P}}(f(\frac{x}{k})).$$

and

$$L_{\mathcal{P}}(f(x)) = \frac{1}{k}L_{k\mathcal{P}}(f(\frac{x}{k}).$$

Proof of Theorem 4.1.1 (iv)

We see that for any partition $\mathcal{P}$ of $[a,b]$,

$$U_{\mathcal{P}}(g(x)) \leq U_{\mathcal{P}}(f(x)).$$

It suffices to take the g.l.b of both sides.

Proof of Theorem 4.1.1 (v)

We simply observe that any $\mathcal{P}$ which is a partition for $[a,b]$ can be refined to a union of a partition $\mathcal{P}_1$ of $[a,c]$ together with a partition $\mathcal{P}_2$ of $[c,b]$ simply by adding the point c. We conclude that

$$L_{\mathcal{P}}(f) \leq L_{\mathcal{P}_1}(f) + L_{\mathcal{P}_2}(f) \leq$$

$$U_{\mathcal{P}_1}(f) + U_{\mathcal{P}_2}(f) \leq U_{\mathcal{P}}(f).$$

Exercises for Section 4.1

1. Let f be bounded and integrable on $[a,b]$. Define the partition $\mathcal{P}_N = \{a, a + \frac{b-a}{N}, \ldots, a + \frac{j(b-a)}{N}, \ldots, b\}$, the division of $[b,a]$ into N equally sized intervals. Let $x_j = a + \frac{j(b-a)}{n}$ be the right endpoint of the jth interval, and define the right Riemann sum $R_N(f) = \sum_{j=1}^{N} f(x_j)\frac{b-a}{N}$.
   a) Show that for any N, we have $L_{\mathcal{P}_N}(f) \le R_N(f) \le U_{\mathcal{P}_N}(f)$.
   b) Let $\mathcal{P}$ be any fixed partition of $[a,b]$. Show that $\lim_{N\to\infty} U_{\mathcal{P}_N}(f) \le U_{\mathcal{P}}(f)$, and $\lim_{N\to\infty} L_{\mathcal{P}_N}(f) \ge L_{\mathcal{P}}(f)$.
   c) Conclude that $\lim_{N\to\infty} L_{\mathcal{P}_N}(f)$ and $\lim_{N\to\infty} U_{\mathcal{P}_N}(f)$ both converge to $\int_a^b f(x)dx$.
   d) Use the squeeze theorem to see that $\lim_{N\to\infty} R_N(f) = \int_a^b f(x)dx$.

2. Prove directly from the definition of the Riemann integral that $\int_0^1 x^2 dx = \frac{1}{3}$. Hint: Partition $[0,1]$ into N equally spaced intervals and use what you know from section 1.1 about the sum of the first N squares. Obtain a lower bound on the lower integral and an upper bound on the upper integral.

3. A function f is *uniformly continuous* on a set A of reals if for every $\epsilon > 0$ there is a $\delta > 0$ so that if $x,y \in A$ with $|x - y| < \delta$, one has $|f(x) - f(y)| < \epsilon$. Let f be a continuous function on the interval $[0,1]$. Show that f is uniformly continuous. Hint: Fix x. From the definition of continuity, show that for every $\epsilon > 0$ there is a $\delta(x) > 0$ with $|x - y| < \delta(x)$ implying that $|f(x) - f(y)| < \epsilon$. Pick $\delta(x)$ as large as possible. If there is a positive lower bound for all the values of $\delta(x)$, then you are done. Otherwise, there is a sequence $\{\delta(x_j)\}$ tending to 0. Pick a subsequence of $\{x_j\}$ tending to a point x (using the Bolzano-Weierstrass theorem). Show that f is not continuous at x.

4. Let f be a continuous function on $[0,1]$. In light of the previous exercise, it is also uniformly continuous. Let $\mathcal{P}_N = \{0, \frac{1}{N}, \ldots, \frac{N-1}{N}, 1\}$. Show that

$$\lim_{N\to\infty} L_{\mathcal{P}_N}(f) = \lim_{N\to\infty} U_{\mathcal{P}_N}(f) =$$

$$\int_0^1 f(x)dx.$$

5. Use the definition of the integral to calculate

$$\int_1^3 x^2 \, dx.$$

Do not use the fundamental theorem of calculus. Hint: Find upper and lower sums arbitrarily close to the integral. You might need the formula

$$\sum_{j=1}^n j^2 = \frac{n(n+1)(2n+1)}{6}.$$

6. Use the definition of the integral to calculate

$$\int_1^2 e^x \, dx.$$

Do not use the fundamental theorem of calculus. Hint: Use the formula for the sum of a finite geometric series. You might need to compute something like

$$\lim_{n \to \infty} \frac{\frac{1}{n}}{e^{\frac{1}{n}} - 1}.$$

You can either use the definition of e or the Taylor series expansion for $e^{\frac{1}{n}}$. They are basically the same thing.

Flipped recitation, Riemann integral

In this section we developed the Riemann integral. You may use a calculator since in each problem you will calculate an upper Riemann sum. The function, interval, and partition are given.

Problem 1 ($I = [1,2]$, $\mathcal{P} = \{1,1.2,1.4,1.6,1.8,2\}$)

$$f(x) = x^4 - -3x^2 + 2.$$

Problem 2 ($I = [0,1]$, $\mathcal{P} = \{0,\frac{1}{n},\frac{2}{n},\ldots,\frac{n-1}{n},1\}$)

$$f(x) = e^x.$$

Problem 3 ($I = [0,1]$, $\mathcal{P} = \{0,\frac{1}{n},\frac{2}{n},\ldots,\frac{n-1}{n},1\}$)

$$f(x) = x^2.$$

◇ 4.2 Integration and uniform continuity

We could write the definition of continuity as follows:

Continuity   A function f is continuous at x if for every $\epsilon > 0$ there exists a $\delta > 0$ so that if $|x - y| < \delta$, then $|f(x) - f(y)| < \epsilon$.

(This is the same as saying that

$$\lim_{y \longrightarrow x} f(y) = f(x),$$

which is how we originally defined continuity.)

One weakness of the above definition (as something that can be applied) is that the number δ depends not just on ϵ and f but also on x. When we try to prove that a function is integrable, we want to control the difference between upper and lower sums. To do this, it would help to have the same δ for a given ϵ work at all choices of x in a particular interval. With this in mind, we make a new definition.

Uniform continuity   We say a function f on the interval $[a,b]$ is *uniformly continuous* if for every $\epsilon > 0$, there is $\delta > 0$ so that whenever $|x - y| < \delta$, we have that $|f(x) - f(y)| < \epsilon$.

The definition of uniform continuity looks very similar to the definition of continuity. The difference is that in the uniform definition, the point x is not fixed. Thus uniform continuity is a stronger requirement than continuity. We now see why uniform continuity is useful for integration.

Theorem 4.2.1   A function f on $[a,b]$ which is uniformly continuous is Riemann integrable on $[a,b]$.

Proof of
Theorem 4.2.1

For every $\epsilon > 0$, there is $\delta > 0$ so that when $x,y \in [a,b]$ and $|x-y| < \delta$, then $|f(x)-f(y)| < \epsilon$.

We now relate the inequality to Riemann sums. Let $I \subset [a,b]$ be an interval of length less than δ. We can see that

$$l.u.b._{x \in I} f(x) - g.l.b._{x \in I} f(x) < \epsilon.$$

Therefore, for any partition $\mathcal{P}$ of $[a,b]$ all of whose intervals are shorter than δ, we have

$$U_{\mathcal{P}}(f) - L_{\mathcal{P}}(f) < \epsilon(b - a).$$

Since this is true for every $\epsilon > 0$, it means that

$$I_{u,[a,b]}(f) = I_{l,[a,b]}(f).$$

So f is integrable on $[a,b]$.

When does it happen that f is uniformly continuous on $[a,b]$? Here's an easy if restrictive condition.

Proposition
4.2.1

Let $f(x)$ be continuous on $[a,b]$ and differentiable at every point of $[a,b]$. Suppose that f' is continuous on $[a,b]$. Then f is uniformly continuous on $[a,b]$.

Proof of
Proposition
4.2.1

Since $f'(x)$ is continuous on $[a,b]$, so is its absolute value $|f'(x)|$ and we may use the extreme value theorem to see that there is a number K so that

$$|f'(x)| \le K$$

for every x in the interval $[a,b]$.

Proof of Proposition 4.2.1 cont. Now choose $\delta = \frac{\epsilon}{K}$. Suppose $x,y \in [a,b]$ with $x < y$ and $y - x < \delta$. We apply the mean value theorem on $[x,y]$ to obtain

$$|f(x) - f(y)|$$
$$= |f'(c)||x - y|$$
$$< \delta K = \epsilon,$$

where $c \in [x,y]$. Thus we shown uniform continuity.

What is nice about Proposition 4.2.1 is that δ depends just on ϵ and the maximum of the derivative. This gives an easy way to predict, say, how many pieces we have to partition an interval into to get a Riemann sum giving a good approximation for the integral. It's really a very quantitative result. However, if all we want is to know whether f is integrable, all this information about the derivative is overkill.

Theorem 4.2.2 Let f be continuous on $[a,b]$. Then f is uniformly continuous on $[a,b]$.

Proof of Theorem 4.2.2 Fix $\epsilon > 0$. For every x, let $\delta(x)$ be the largest δ that works for ϵ and x in the definition of continuity. More precisely:

$$\delta(x) = l.u.b.\{\delta : \delta \leq b - a \text{ and}$$

$$\text{when } |x - y| < \delta, |f(y) - f(x)| < \epsilon\}$$

The function $\delta(x) > 0$. Also, the function $\delta(x)$ is continuous (why? An exercise for the reader! It's basically Exercise 4.1.3) By the extreme value theorem, this means that $\delta(x)$ has a minimum on $[a,b]$. Hence, f is uniformly continuous.

What we get from this is that every continuous function on a closed interval is Riemann integrable on the interval. That's a lot of functions. But, in fact, many more functions are integrable. For instance, a function f on $[a,b]$ having finitely many points of discontinuity at which all the left and right limits exist and are finite is also integrable. You see this by restricting to partitions containing the points of discontinuity. An exhaustive description of all Riemann integrable function is (slightly) beyond the scope of this book.

Exercises for Section 4.2

1. Let f be a function on the reals whose derivative f' is defined and continuous on $[a,b]$. Let $\mathcal{P}_N$ and $R_N(f)$ be as in exercise 4.1.1.

    a) Observe using the extreme value theorem that there is a real number M so that $|f'(x)| \leq M$ for every x in $[a,b]$.

    b) See using the mean value theorem that if $x,y \in [a,b]$ with $|x - y| \leq \frac{\epsilon}{M}$, where M is the constant in part a), that $|f(x) - f(y)| \leq \epsilon$.

    c) Show directly using part b) that $|R_N(f) - U_{\mathcal{P}_N}(f)| \leq \frac{M(b-a)^2}{N}$. Similarly show that $|R_N(f) - L_{\mathcal{P}_N}(f)| \leq \frac{M(b-a)^2}{N}$.

    d) Conclude using part c) that $|R_N(f) - \int_a^b f(x)dx| \leq \frac{M(b-a)^2}{N}$.

2. Suppose f is a bounded function on the interval $[a,b]$. Show that f is integrable on $[a,b]$ if and only if for each $\epsilon > 0$, there is a partition $\mathcal{P}$ of $[a,b]$ so that

$$U(f,\mathcal{P}) - L(f,\mathcal{P}) < \epsilon.$$

3. Let f be a bounded function on $[a,b]$. For each $x \in (a,b)$, define $o(f,x)$, the *oscillation* of f at x, as

$$\lim_{\delta \longrightarrow 0} (l.u.b_{y\in(x-\delta,x+\delta)} f(y) -$$

$$g.l.b_{y\in(x-\delta,x+\delta)} f(y)).$$

Show that $o(f,x)$ always exists and that f is continuous at x if and only if $o(f,x) = 0$.

4. Let f and $o(f,x)$ be as in exercise 3. Fix $\epsilon > 0$ Suppose there are a finite set of intervals $I_1,\ldots,I_n$ of total length at most ϵ so that every x with $o(f,x) > \epsilon$ is contained in one of the I_j. Then show there is a partition $\mathcal{P}$ of $[a,b]$ so that

$$U(f,\mathcal{P}) - L(f,\mathcal{P}) = O(\epsilon),$$

where the implicit constants depend on a,b and the maximum and minimum values of f but nothing else.

Flipped recitation, uniform continuity

In this section we developed the notion of uniform continuity. In each problem, you are given a closed interval and a continuous function on it. The function is therefore uniformly continuous. Give a choice of $\delta(\epsilon)$ for each $\epsilon > 0$, which satisfies the definition of uniform continuity for the function on the interval?

Problem 1 $(I = [1,5])$

$$f(x) = e^x.$$

Problem 2 $(I = [-2,2])$

$$f(x) = x^2.$$

Problem 3 $(I = [0,1])$

$$f(x) = \sqrt{x}.$$

◇ **4.3   The fundamental theorem**

Differentiation and integration are not unrelated, which is why calculus is a unified subject. As their names suggest, in a certain sense they are opposites. We will be precise about what sense in what follows as we state and prove two forms of the fundamental theorem of calculus.

Theorem 4.3.1 Fundamental theorem of Calculus, version 1

Let F be a continuous function on the interval $[a,b]$. Suppose F is differentiable everywhere in the interior of the interval with derivative of f which is Riemann integrable. Then

$$\int_a^b f(x)dx = F(b) - F(a).$$

Proof of fundamental theorem of calculus, version 1

Let

$$\mathcal{P} = \{x_0, x_1, \ldots, x_n\}$$

be any partition of $[a,b]$. We now apply the mean value theorem to F on each subinterval $[x_{j-1}, x_j]$. We conclude that there is $c_j \in (x_{j-1}, x_j)$ so that

$$F(x_j) - F(x_{j-1}) = f'(c_j)(x_j - x_{j-1}).$$

Now we sum over j. We obtain

$$\sum_{j=1}^n F(x_j) - F(x_{j-1}) = \sum_{j=1}^n f'(c_j)(x_j - x_{j-1}).$$

Proof of fundamental theorem of calculus, version 1 cont.

The key points here are that the left-hand side telescopes and the right-hand side is a Riemann sum (though probably neither upper nor lower). Thus

$$F(b) - F(a) = \sum_{j=1}^{n} f'(c_j)(x_j - x_{j-1}),$$

from which we conclude that

$$L_{\mathcal{P}}(f) \leq F(b) - F(a) \leq U_{\mathcal{P}}(f).$$

Since we assumed that f is integrable, we can obtain $\int_a^b f(x)dx$ both as $l.u.bL_{\mathcal{P}}(f)$ and as $g.l.b.U_{\mathcal{P}}(f)$. Thus we obtain

$$\int_a^b f(x)dx \leq F(b) - F(a) \leq \int_a^b f(x)dx.$$

Therefore

$$\int_a^b f(x)dx = F(b) - F(a),$$

as desired.

Theorem 4.3.2 fundamental theorem of calculus, version 2

Let f be continous on $[a,b]$ and let

$$F(x) = \int_a^x f(y)dy.$$

Then $F'(x) = f(x)$.

Proof of fundamental theorem of calculus, version 2

We calculate

$$F'(x) = \lim_{h \longrightarrow 0} \frac{1}{h} \int_x^{x+h} f(y)dy.$$

Since f is continuous, for any $\epsilon > 0$ there exists $\delta > 0$ so that if $|y - x| < \delta$, then we have $|f(y) - f(x)| < \epsilon$. Thus if we pick $h < \delta$, we have

$$\left| \int_x^{x+h} f(y)dy - hf(x) \right| < \int_x^{x+h} \epsilon dy = \epsilon h.$$

Thus

$$\left| \frac{1}{h} \int_x^{x+h} f(y)du - f(x) \right| < \epsilon.$$

We have shown that the limit in h converges to $f(x)$.

An immediate consequence is that knowing the value of a sufficiently nice function at one point and knowing its derivative everywhere is enough to define the function.

Corollary 4.3.1

Let F be a function which has a continuous derivative f on an interval $[a,b]$. Then

$$F(x) = F(a) + \int_a^x f(y)dy.$$

Another application of the fundamental theorem is that we may apply rules for differentiation to integration. We use the chain rule to obtain the change of variables formula.

Theorem 4.3.3 change of variables formula   Let f be integrable on an interval $[a,b]$. Let $g(x)$ be a differentiable function taking the interval $[c,d]$ to the interval $[a,b]$ with $g(c) = a$ and $g(d) = b$. Then

$$\int_a^b f(x)dx = \int_c^d f(g(x))g'(x)dx.$$

Proof of Theorem 4.3.3   Let F be the antiderivative of f. We may obtain F by

$$F(x) = \int_a^x f(y)dy.$$

Then clearly

$$\int_a^b f(x)dx = F(b) - F(a).$$

On the other hand, clearly by the chain rule, $f(g(x))g'(x)$ is the derivative of $F(g(x))$. Thus

$$\int_c^d f(g(x))g'(x)dx = F(g(d)) - F(g(c))$$

$$= F(b) - F(a).$$

We have shown that the two integrals are equal.

In a high school calculus course, the change of variables formula is usually presented as substitution. We substitute $u = g(x)$. Then $du = g'(x)dx$. And we can get from one integral to another. In effect, the change of variables formula justifies the notation dx. We don't know what the differential is, but it follows the right rules. You will hear more along these lines when studying multivariable calculus.

Another application of a rule of differentiation to integration is that the product rule becomes integration by parts. The challenge, of course, is deciding how to factor the integrand as a product.

Theorem 4.3.4 Integration by parts	Let f and g be once continuously differentiable functions on $[a,b]$. Then $$\int_a^b f(x)g'(x)dx =$$ $$-\int_a^b f'(x)g(x) + f(b)g(b) - f(a)g(a).$$

We end this section by discussing improper integrals and an application. If f is bounded and integrable on all intervals of nonnegative reals, we can define

$$\int_0^\infty f(x)dx = \lim_{y \to \infty} \int_0^y f(x)dx.$$

Similarly, if f is bounded and integrable on all intervals $[a,y]$ with $y < b$, we can define

$$\int_a^b f(x)dx = \lim_{y \to b} \int_a^y f(x)dx.$$

These integrals are called *improper* and only converge if the limit defining them converges.

As an application of the notion of improper integrals, we obtain a version of the integral test for convergence of series.

**Theorem
4.3.5**

Let f be a decreasing, nonnegative function of the positive reals. Then

$$\sum_{j=1}^{\infty} f(j)$$

converges if

$$\int_1^{\infty} f(x)dx$$

converges. Further,

$$\sum_{j=1}^{\infty} f(j)$$

diverges if

$$\int_1^{\infty} f(x)dx$$

diverges.

**Proof of
Theorem 4.3.5**

Simply observe that the series is bounded below by $\int_1^{\infty} f(x)dx$ and that the series $\sum_{j=2}^{\infty} f(j)$ is bounded above by the same integral.

As an example, we can consider series like $\sum \frac{1}{n \log n}$ and $\sum \frac{1}{n(\log n)^2}$. We haven't specified the starting point of the sum so that there are no zeroes in the denominator.

We should compare to

$$\int \frac{dx}{x \log x}$$

and

$$\int \frac{dx}{x(\log x)^2}.$$

We make the substitution $u = \log x$ and $du = \frac{dx}{x}$. Then the integrals become

$$\int \frac{du}{u},$$

and

$$\int \frac{du}{u^2}.$$

How do these substitutions compare to the term grouping arguments made when we studied the same sums previously?

Exercises for Section 4.3

1. Let $f(x)$ be a continuous function on $[a,b]$. Define $F(x)$ on $[a,b]$ by $F(x) = \int_a^x f(y)dy$. Prove for any $c \in (a,b)$ that $F'(c) = f(c)$. Hint: Calculate the limit directly. Use the continuity of f at c.

2. Prove that

$$\sum_{n=2021}^{\infty} \frac{1}{n \log n \log(\log n)}$$

diverges.

3. Prove Theorem 4.3.4.

4. Let $g(x)$ be a function which is continuous on the whole real line. Let $k(x,t)$ be a function of two variables so that for each fixed value of x, the function $k(x,t)$ is continuous as a function of t on the whole real line. Suppose further that for each fixed value of t, the function $k(x,t)$ is continuously differentiable as a function of x. Show that

$$f(x) = \int_0^x k(x,t)g(t)dt$$

is defined at every positive real x. Show that f is in fact differentiable at every positive real x by writing down and proving a formula for $f'(x)$. (Hint: If in doubt, write out the derivative as a limit.)

Flipped recitation, fundamental theorem of Calculus

In this section we proved the fundamental theorem of calculus

Problem 1 Calculate

$$\int_2^3 \frac{dx}{x(x-1)}.$$

Problem 2 Calculate

$$\int_5^6 x^2 e^x dx.$$

Problem 3 Calculate

$$\int_6^7 \frac{2x^2 dx}{\sqrt{x^3+7}}.$$

◇ 4.4   Taylor's theorem with remainder

In Chapter 3, we proved the weak Taylor approximation, Theorem 3.4.4.

Theorem 4.4.1 Weak Taylor again

Let f be a function on an interval $[a,b]$ and c a point in the interior of the interval. Suppose that f is $n-2$ times continuously differentiable on $[a,b]$, that the $n-1$st derivative of f exists everywhere on the interior of (a,b), and that the nth derivative of f exists at c. Then

$$f(x) = f(c) + f'(c)(x-c) + \cdots + \frac{f^{(n)}(c)}{n!}(x-c)^n$$

$$+ o((x-c)^n).$$

The expression $f(c) + f'(c)(x-c) + \cdots + \frac{f^{(n)}(c)}{n!}(x-c)^n$ is referred to as the nth *Taylor polynomial* of f at c. It is, of course, a polynomial of degree n which approximates f at c. We know that the error is $o((x-c)^n)$, so it is getting small quite fast as x approaches c. But, of course, the definition of $o(.)$ involves a limit, and we don't know how fast that limit converges. We will rectify this now, obtaining a more concrete estimate of the error in good cases. We will need to use integration by parts.

Lemma 4.4.1

Let f and g be once continuously differentiable functions on the interval $[a,b]$. Then

$$\int_a^b (f'(x)g(x) + f(x)g'(x))dx$$

$$= f(b)g(b) - f(a)g(a).$$

Proof of Lemma 4.4.1   Just apply the product rule to convert the integral to

$$\int_a^b \frac{d}{dx}(f(x)g(x))dx,$$

and apply the fundamental theorem of calculus.

Theorem 4.4.2 Taylor's theorem with remainder, part 1   Assume f is $n+1$ times continuously differentiable in the interval $[a,b]$ having c in the interior. Then for every $x \in [a,b]$ we have

$$f(x) = \sum_{j=0}^n \frac{f^{(j)}(c)}{j!}(x-c)^j + R_n(x),$$

with

$$R_n(x) = \frac{1}{n!}\int_c^x (x-y)^n f^{(n+1)}(y)dy.$$

In Theorem 4.4.2, the expression $R_n(x)$ is referred as the *remainder* in the nth Taylor approximation of f at a.

Proof of Taylor's theorem with remainder, part 1

We will prove this, of course, by induction. The base case, $n = 0$, is nothing more than the fundamental theorem of calculus, so we will assume that

$$R_n(x) = \frac{1}{n!} \int_c^x (x - y)^n f^{(n+1)}(y)\,dy,$$

and we will try to calculate R_{n+1} under the assumption that f has $n+2$ continuous derivatives. We observe that as long as the result holds

$$R_{n+1}(x) = R_n(x) - \frac{f^{(n+1)}(c)}{(n+1)!}(x - c)^{n+1}.$$

Now we combine this with the induction hypothesis, taking the $n + 1$ factor from the denominator to turn $(x - c)^{n+1}$ into an integral from c to x; namely, $\frac{(x-c)^{n+1}}{n+1} = \int_c^x (x - y)^n dy$

$$R_{n+1}(x)$$
$$= \frac{1}{n!} \int_c^x (x - y)^n f^{(n+1)}(y)\,dy$$
$$- \frac{f^{(n+1)}(c)}{n!} \int_c^x (x - y)^n dy$$
$$= \frac{1}{n!} \int_c^x (x - y)^n (f^{(n+1)}(y) - f^{(n+1)}(c))\,dy$$
$$= \frac{1}{(n+1)!} \int_c^x (x - y)^{n+1} f^{(n+2)}(y)\,dy.$$

Here, the last step is by integration by parts. We integrate $(x - y)^n$ and differentiate $f^{(n+1)}(y) - f^{(n+1)}(c)$. Note that the boundary terms vanish since $(x - y)^n$ vanishes at $y = x$ and $f^{(n+1)}(y) - f^{(n+1)}(c)$ vanishes at $y = c$.

Having this remarkable formula for R_n, we look for a way to apply it. We first write down a general result about integrals of continuous functions which is in analogy with the mean value theorem.

Theorem 4.4.3 Mean value theorem for integrals

Let f and g be continuous functions on $[a,b]$. Assume that g never changes sign on $[a,b]$. Then there is $c \in (a,b)$ so that

$$\int_a^b f(x)g(x)dx = f(c) \int_a^b g(x)dx.$$

Proof of mean value theorem For integrals

Let

$$M = \frac{\int_a^b f(x)g(x)dx}{\int_a^b g(x)dx}.$$

Suppose that everywhere on (a,b), we have that $f(x) < M$. Then

$$\int_a^b f(x)g(x)dx < M \int_a^b g(x)dx,$$

which is a contradiction. Similarly, it is not the case that $f(x) > M$ for every x in (a,b). Then since f is continuous, by the intermediate value theorem, there must be c so that $f(c) = M$.

**Theorem
4.4.4
Taylor's
theorem
with
remainder,
part 2**

Assume f is $n+1$ times continuously differentiable in the interval $[a,b]$ having c in the interior. Then for every $x \in [a,b]$ we have

$$f(x) = \sum_{j=0}^{n} \frac{f^{(j)}(c)}{j!}(x-c)^j + R_n(x).$$

and there exists a number d strictly between x and c so that

$$R_n(x) = \frac{f^{(n+1)}(d)}{(n+1)!}(x-c)^{n+1}.$$

**Proof of
Taylor's
theorem with
remainder,
part 2**

We now apply this mean value theorem for integrals to our expression for the remainder in Taylor's approximation.

$$R_n(x) = \frac{1}{n!} \int_c^x (x-y)^n f^{(n+1)}(y)dy.$$

Observe that

$$\frac{1}{n!} \int_c^x (x-y)^n dy = \frac{1}{(n+1)!}(x-c)^{n+1}.$$

Thus we observe that there is some d between c and x so that

$$R_n(x) = \frac{f^{(n+1)}(d)(x-c)^{n+1}}{(n+1)!}.$$

We conclude that

$$f(x) = \sum_{j=0}^{n} \frac{f^{(j)}(c)}{j!}(x-c)^j + \frac{f^{(n+1)}(d)(x-c)^{n+1}}{(n+1)!}.$$

If we begin with an *a priori* estimate for the $(n+1)$st derivative, Taylor's theorem with remainder gives us an explicit estimate for the error term.

Exercises for Section 4.4

1. Let $I = \int_0^1 \frac{1+x^{2014}}{1+x^{10000}} dx$. Show that $I = 1 + \frac{c}{2015}$ for some $0 < c < 1$. Hint: Observe that $1 < \frac{1+x^{2014}}{1+x^{10000}} < 1 + x^{2014}$ whenever $0 < x \leq 1$.

2. Let $0 < x < 1$. Let $f(x) = \frac{1}{1-x}$. Write down an explicit expression for $R_2(x)$ as a rational function. Show directly from this expression that $R_2(x) = \frac{1}{6} f^{(3)}(d) x^3$ for some $0 < d < x$. Solve for d in terms of x.

Flipped recitation, Taylor's theorem with remainder

In the section, we developed Taylor's theorem with remainder. For each function given below, calculate the Taylor series up to order n specified at 0, evaluated at 1 and calculate the maximum of the $n+1$st derivative on $[0,1]$ to obtain a bound for the error between what you have calculated and the function evaluated at 1.

Problem 1 $n = 5$

$$f(x) = e^x.$$

Problem 2 $n = 10$

$$f(x) = e^x.$$

Problem 3 $n = 4$

$$f(x) = \sqrt{9 + x}.$$

◇ 4.5   Numerical integration

We have developed the fundamental theorem of calculus which allows us to calculate explicitly the integrals of traditional calculus. The idea is that there is a library of functions, the elementary functions which we should feel free about using, and as long as a function has an elementary function as an antiderivative, we may compute its integral. An example is the function e^{-x^2}. It is viewed as elementary because e^x is elementary and $-x^2$ is elementary and the class of elementary functions is closed under composition.

Now the function e^{-x^2} certainly has an antiderivative. Because e^{-x^2} is continuous on any closed interval, it is uniformly continuous on the interval and so it is integrable. Thus for any y, we may define

$$\int_0^y e^{-x^2}\,dx.$$

By simply setting

$$F(y) = \int_0^y e^{-x^2}\,dx,$$

we have defined an antiderivative for e^{-x^2}. This function is important. In fact, it shows up in many basic applications in probability. But this function $F(y)$ is not elementary, in the sense that we can't build it up by addition, multiplication, and composition from our library of simpler functions and we don't *a priori* have a good approach to computing it except through the definition of integration. We can write down a lower sum and an upper sum for the integral with respect to some partition and this gives us a range in which the value of the function is contained.

The purpose of this section is to study and compare different methods of approximating the integral

$$\int_a^b f(x)\,dx.$$

Given a method of approximation, our goal will be to study how big the error is. We will make some assumptions about our function f. It will be infinitely continuously differentiable,

but for each method of integration, we will only make use of a
few derivatives. We will assume that all the derivatives we use
are bounded by a constant M on the interval $[a,b]$. (When we
change the number of derivatives we use for a given function,
we may have to give something up in the constant M.) We
now describe various methods of numerical integration.

**The sloppy
method**
We partition our interval $[a,b]$ into equally
spaced subintervals using the partition

$$P = \{a, a + \frac{(b-a)}{n}, a + \frac{2(b-a)}{n}, \ldots, a + \frac{j(b-a)}{n}$$

$$, \ldots, a + \frac{(n-1)(b-a)}{n}, b\}.$$

Now for each subinterval $I_j = [a + \frac{(j-1)(b-a)}{n}, a + \frac{j(b-a)}{n}]$, we pick a point $c_j \in I_j$.
We evaluate

$$J_1 = \sum_{j=1}^{n} (\frac{(b-a)}{n}) f(c_j).$$

The quantity J_1 given by the sloppy method is a Riemann
sum corresponding to the choice of points c_j. What can we say
a priori about how close the quantity J_1 is to the the integral?

We let $x \in I_j$. We will use only the first derivative of f and
assume that it is everywhere less than M. By the mean value
theorem, we know that

$$|f(x) - f(c_j)| \le M|x - c_j| \le \frac{M(b-a)}{n}.$$

Thus we estimate that

$$|\int_{a + \frac{(j-1)(b-a)}{n}}^{a + \frac{j(b-a)}{n}} f(x)dx - f(c_j)\frac{(b-a)}{n}| \le \frac{M(b-a)^2}{n^2}.$$

Using the triangle inequality and the fact we have subdivided

$[a,b]$ into n intervals, we get

$$|\int_a^b f(x)dx - J_1| \le \frac{M(b-a)^2}{n}.$$

The key thing to take away about this estimate on the error $|\int_a^b f(x)dx - J_1|$ is that it is $O(\frac{1}{n})$. Most of our work is in evaluating the function n times and we obtain an error estimate for the integral which is $O(\frac{1}{n})$. Can we do any better?

From this point on, we will give up on explicit estimates in terms of M in favor of asymptotic estimates.

The midpoint method The midpoint method is the very best version of the sloppy method. Here we take c_j to be the midpoint of I_j, namely,

$$m_j = a + \frac{(j - \frac{1}{2})(b-a)}{n}.$$

We define

$$J_2 = \sum_{j=1}^{n}(\frac{(b-a)}{n})f(m_j).$$

The key point will be that for any $x \in I_j$, we have the estimate

$$f(x) = f(m_j) + f'(m_j)(x - m_j) + O((x - m_j)^2)$$

by, for instance, Taylor's approximation to order 2. (This is a key point. We're using two derivatives of the function instead of one, which is all we used in the sloppy method.) Next we observe that

$$\int_{a+\frac{(j-1)(b-a)}{n}}^{a+\frac{j(b-a)}{n}} (f(m_j) + f'(m_j)(x - m_j))dx = f(m_j)(\frac{b-a}{n}).$$

Thus

$$|\int_{a+\frac{(j-1)(b-a)}{n}}^{a+\frac{j(b-a)}{n}} f(x)dx - f(m_j)(\frac{b-a}{n})| = \frac{b-a}{n}O((\frac{b-a}{n})^2).$$

Thus we obtain after summing in n that

$$|J_2 - \int_a^b f(x)dx| = O(\frac{1}{n^2}).$$

It pays to pick the midpoint.

Simpson's rule Here is a method of integration of a slightly different type that beats the midpoint rule. To simplify our notation we let

$$x_j = a + \frac{j(b-a)}{n},$$

be the typical point of the partition. As before, we let

$$m_j = \frac{x_{j-1} + x_j}{2}.$$

We let

$$J_3 = \sum_{j=1}^n \frac{f(x_{j-1}) + 4f(m_j) + f(x_j)}{6}(\frac{b-a}{n}).$$

In trying to understand why J_3 is a good estimate for the integral $\int_a^b f(x)dx$, we make the following observation.

Proposition 4.5.1 Let $I = [x_l, x_r]$ be any interval and let $q(x)$ be a quadratic polynomial on I. Let x_m be the midpoint of I. Then

$$\int_{x_l}^{x_r} q(x)dx = \frac{x_r - x_l}{6}(q(x_l) + 4q(x_m) + q(x_r)).$$

Proof **of** It suffices to prove this for the interval $[-h,h]$,
Proposition since we can move any interval to this by shift-
4.5.1 ing the midpoint to 0 (and that operation pre-
serves quadraticness). We calculate

$$\int_{-h}^{h} (cx^2 + dx + e)dx = \frac{2}{3}ch^3 + 2eh.$$

We observe that this is exactly what the
proposition predicts.

Using the claim, we invoke the fact that on each interval
I_j, we have

$$f(x) = f(m_j) + (x-m_J)f'(m_j) + (x-m_j)^2\frac{f''(m_j)}{2} + O((x-m_j)^3).$$

That is, each function that is three times differentiable has
a good quadratic approximation on each integral. We use the
claim to bound the difference between the jth term in the
sum for J_3 and the part of the integral from x_{j-1} to x_j by
$\frac{b-a}{n}O((\frac{1}{n})^3)$. We conclude, at last, that

$$\left|J_3 - \int_a^b f(x)dx\right| = O(\frac{1}{n^3}).$$

In the exercises, we'll see we can do a bit better.

Exercises for Section 4.5

1. Define the trapezoid rule as follows. With notation as in the text, let

$$J_4 = \sum_{j=1}^{n} \frac{f(x_{j-1}) + f(x_j)}{2} \frac{b-a}{n}.$$

a) Let $f(x)$ be a quadratic function, $f(x) = cx^2 + dx + e$. Give an explicit expression for the difference between J_4 and $\int_a^b f(x)dx$.
b) Do the same analysis with a quadratic f for J_2, the error obtained from the midpoint rule.
c) Compare a) and b). Derive Simpson's rule as an exact method of integrating quadratics.

2. Apply Simpson's rule to a cubic function $f(x) = cx^3 + dx^2 + ex + f$. Calculate an explicit expression for the error. What can you conclude about the accuracy of Simpson's rule for general functions?

3. Suppose you are given that a function $q(x)$ on the interval $[10,22]$ satisfies

$$q(10) = 1, \ q(13) = 3, \ q(16)$$
$$= 7, \ q(19) = 15, \ q(22) = 20.$$

Suppose you are given further that $q(x)$ is a polynomial of degree 4 (also called a quartic polynomial). Write down explicitly the Taylor series for q centered at 16.

4. Find a formula giving the integral $\int_{-1}^{1} q(x)dx$ of a general quartic polynomial $q(x) = ax^4 + bx^3 + cx^2 + dx + e$ in terms of the values of q at $-1, \frac{-1}{2}, 0, \frac{1}{2}, 1$. Use this to define a numerical integration method for functions with six continuous derivatives. What is the order of the error for this method of differentiation?

Flipped recitation, numerical methods of integration

In this section, we developed numerical methods of integration. In each problem approximate

$$\int_0^1 f(x)dx$$

using the specified method and number of steps.

Problem 1 Sloppy method

$$f(x) = x^2 + e^x \quad n = 20$$

Problem 2 Midpoint rule

$$f(x) = x^2 + e^x \quad n = 10$$

Problem 3 Simpson's rule

$$f(x) = x^2 + e^x \quad n = 5$$

Chapter 5

CONVEXITY

◇ 5.1 Convexity and optimization

We say that if f is a once continuously differentiable function on an interval I, and x is a point in the interior of I, x is a *critical point* of f if

$$f'(x) = 0.$$

Critical points of once continuously differentiable functions are important because they are the only points that can be local maxima or minima.

In the context of critical points, the second derivative of a function f is important because it helps in determining whether a critical point is a local maximum or minimum. We recall the first and second derivative tests from Section 3.3, theorem 3.3.2 and 3.3.3:

Theorem 5.1.1 First derivative test, again	Let f be a once continuously differentiable function on an interval I and let x be a critical point. Suppose there is some open interval (a,b) containing x so that for $y \in (a,b)$ with $y < x$, we have $f'(x) > 0$, and so that for $y \in (a,b)$ with $y > x$, we have $f'(x) < 0$. Then f has a local maximum at x. If on the other hand, we have $f'(y) < 0$ when $y \in (a,b)$ and $y < x$ and $f'(y) < 0$ when $y \in (a,b)$ and $y > x$, then f has a local minimum at x.

Note that under these assumptions, we actually have that x is the unique global maximum (or minimum) on the interval [a,b]. Intuitively, this says we get a maximum if f is increasing as we approach x from the left and if f is decreasing as we leave x to the right.

Theorem 5.1.2 Second derivative test, again   Let f be a once continuously differentiable function. Let x be a critical point for f. If f is twice differentiable at x and $f''(x) < 0$, then f has a local maximum at x. If f is twice differentiable at x and $f''(x) > 0$, then f has a local minimum at x.

All of this is closely related to the notion of convexity and concavity.

Definition of concavity and convexity   A function $f(x)$ is concave if it lies above all its secants. Precisely, f is *concave* if for any a,b,x with $x \in (a,b)$, we have

$$f(x) \geq \frac{b-x}{b-a}f(a) + \frac{x-a}{b-a}f(b).$$

We say f is *strictly concave* if under the same conditions

$$f(x) > \frac{b-x}{b-a}f(a) + \frac{x-a}{b-a}f(b).$$

Similarly, f is convex if it lies below all its secants. Precisely, f is *convex* if for any a,b,x with $x \in (a,b)$, we have

$$f(x) \leq \frac{b-x}{b-a}f(a) + \frac{x-a}{b-a}f(b).$$

We say f is *strictly convex* if under the same conditions

$$f(x) < \frac{b-x}{b-a}f(a) + \frac{x-a}{b-a}f(b).$$

Theorem 5.1.3

Let f be twice continuously differentiable. Then f is concave if and only if for every x, we have $f''(x) \le 0$, and it is convex if and only if for every x, we have $f''(x) \ge 0$.

We will leave the proof of Theorem 5.1.3 as an exercise but we indicate briefly why this is true locally. If $f''(x) \le 0$, we have

$$f(y) = f(x) + f'(x)(y - x) + \frac{f''(x)}{2}(y - x)^2 + o((y - x)^2).$$

We observe that for a,b close to x, if $f''(x) < 0$, we have that $(a,f(a))$ and $(b,f(b))$ are below the tangent line to the graph of f at x. Thus the point $(x,f(x))$ which is on the tangent line is above the secant between $(a,f(a))$ and $(b,f(b))$.

Concavity has a lot to do with optimization.

Example 1   Resource allocation problems.

Let f and g be two functions which are continuous on $[0,\infty)$ and twice continuously differentiable with a strictly negative second derivative on $(0,\infty)$. Let t be a fixed number.

Consider

$$W(x) = f(x) + g(t - x).$$

(Interpretation: You have t units of a resource and must allocate them between two productive uses. The function of W might represent the value of allocating x units to the first use and $t - x$ units to the second use. The concavity represents the fact that each use has diminishing returns).

Under these assumptions, if $W(x)$ has a critical point in $(0,t)$, then the critical point is a maximum.

If, in addition,

$$\lim_{x \to 0} f'(x) = \infty$$

and

$$\lim_{x \to 0} g'(x) = \infty,$$

then we are guaranteed that $W(x)$ has a unique maximum in $[0,t]$. This is because

$$\lim_{x \to 0} W'(x) = \infty$$

and

$$\lim_{x \longrightarrow t} W'(x) = -\infty.$$

By the intermediate value theorem, there is a zero for $W'(x)$ in $(0,t)$. The zero is unique since $W'(x)$ is strictly decreasing.

Strictly concave functions on $(0,\infty)$ whose derivatives converge to ∞ at 0 are ubiquitous in economics. We give an example.

Example 2   Cobb-Douglas production function
The Cobb-Douglas production function gives the output of an economy as a function of its inputs (labor and capital).

$$P(K,L) = cK^{\alpha}L^{1-\alpha}.$$

Here c is a positive constant and α a real number between 0 and 1. The powers of K and L in the function have been chosen so that

$$P(tK,tL) = tP(K,L).$$

That is, if we multiply both the capital and the labor of the economy by t, then we multiply the output by t. Note that if we hold L constant and view $P(K,L)$ as a function of K, then we see this function as defined on $(0,\infty)$ is strictly concave with the derivative going to ∞ at 0.

An important principle of economics is that we should pay for capital at the rate of the marginal product of capital. We find this rate by taking the derivative in K and getting $\alpha cK^{\alpha-1}L$. Since we need K units of capital to get the economy to function, we pay $\alpha cK^{\alpha}L^{1-\alpha}$. In this way, we see that α represents the share of the economy that is paid to the holders of capital and $1 - \alpha$ is the share paid to the providers of labor.

1. Let a function f be twice continuously differentiable on $[a,b]$. Show that f is concave if and only if $f''(x) \leq 0$ for every $x \in (a,b)$.

2. Let f and g be continuous on $[0,\infty)$ and twice continuously differentiable on $(0,\infty)$. Suppose that f and g are increasing and concave on $(0,\infty)$ and suppose that $g(0) = 0$ and $g(t) > 0$ for $t > 0$. Show that

$$h(x) = f(g(x))$$

is increasing and concave.

Flipped recitation, Cobb-Douglas machine
In this section, we developed the relationship between convexity and optimization and we defined the Cobb-Douglas machine. For Cobb-Douglas machines with parameters given below, calculate the price of labor and capital.

Problem 1

$$c = 1 \quad \alpha = \frac{1}{3} \quad K = 5 \quad L = 10$$

Problem 2

$$c = 1 \quad \alpha = \frac{1}{3} \quad K = 9 \quad L = 6$$

Problem 3

$$c = 1 \quad \alpha = \frac{1}{2} \quad K = 5 \quad L = 10$$

◇ 5.2   Inequalities

Another way in which optimization can be applied is to prove inequalities.

Theorem 5.2.1 Arithmetic geometric mean inequality

Let a and b be positive numbers. Then

$$a^{\frac{1}{2}}b^{\frac{1}{2}} \leq \frac{1}{2}(a+b).$$

This can be proved in a purely algebraic way.

Algebraic proof of arithmetic geometric mean inequality

$$a + b - 2a^{\frac{1}{2}}b^{\frac{1}{2}} = (a^{\frac{1}{2}} - b^{\frac{1}{2}})^2.$$

Analytic proof of arithmetic geometric mean inequality

It suffices to prove the inequality when $a+b = 1$. This is because

$$(ta)^{\frac{1}{2}}(tb)^{\frac{1}{2}} = ta^{\frac{1}{2}}b^{\frac{1}{2}}$$

while

$$(ta + tb) = t(a + b),$$

so we just pick $t = \frac{1}{a+b}$.

Analytic proof of arithmetic geometric mean inequality cont.

Thus what we need to prove is

$$\sqrt{x}\sqrt{1-x} \le \frac{1}{2}$$

when $0 < x < 1$. We let

$$f(x) = \sqrt{x}\sqrt{1-x}$$

and calculate

$$f'(x) = \frac{\sqrt{1-x}}{2\sqrt{x}} - \frac{\sqrt{x}}{2\sqrt{1-x}}.$$

$$f''(x) = -\frac{1}{2\sqrt{x}\sqrt{1-x}} - \frac{\sqrt{x}}{4(1-x)^{\frac{3}{2}}} - \frac{\sqrt{1-x}}{4x^{\frac{3}{2}}}.$$

All terms in the last line are negative, so f is strictly concave. The unique critical point is at $x = \frac{1}{2}$, where equality holds. We have shown that

$$\sqrt{x}\sqrt{1-x} \le \frac{1}{2},$$

since $\frac{1}{2}$ is the maximum.

The analytic proof looks a lot messier than the algebraic one, but it is more powerful. For instance, by the same method, we get that if $\alpha, \beta > 0$ and

$$\alpha + \beta = 1,$$

then

$$a^\alpha b^\beta \le \alpha a + \beta b$$

for $a, b > 0$. This simply requires applying concavity for the function

$$f(x) = x^\alpha (1-x)^\beta,$$

and finding the unique maximum.

Closely related to the AGM inequality is the geometric harmonic mean inequality. This is the simplest version:

Theorem 5.2.2
Simple version of Harmonic Geometric mean inequality

Let $a,b > 0$ be real numbers. Then

$$\frac{2}{\frac{1}{a} + \frac{1}{b}} \le a^{\frac{1}{2}} b^{\frac{1}{2}}.$$

Proof of Theorem 5.2.2

Multiply the numerator and the denominator of the left-hand side by ab. Then divide both sides by $a^{\frac{1}{2}} b^{\frac{1}{2}}$ and take the reciprocal and you obtain the arithmetic geometric mean inequality. All steps are reversible, so the two inequalities are equivalent.

In the same way, we can obtain a more general (weighted) version. Let $\alpha, \beta > 0$ with $\alpha + \beta = 1$. Then

$$\frac{1}{\frac{\alpha}{a} + \frac{\beta}{b}} \le a^{\alpha} b^{\beta}.$$

We can obtain similar results for sums of not just two terms but n terms.

Theorem 5.2.3
n-term AGM inequality

Let $\alpha_1, \ldots, \alpha_n > 0$ with

$$\sum_{j=1}^{n} \alpha_j = 1.$$

Then

$$a_1^{\alpha_1} a_2^{\alpha_2} \ldots a_n^{\alpha_n} \le \sum_{j=1}^{n} \alpha_j a_j.$$

Proof of Theorem 5.2.3   We prove this by induction on n. The base case, $n = 2$, is already known. We let

$$\alpha = \sum_{j=1}^{n-1} \alpha_j$$

and $\beta = \alpha_n$. We let

$$a = (a_1^{\alpha_1} \ldots a_{n-1}^{\alpha_{n-1}})^{\frac{1}{\alpha}}$$

and $b = a_n$. Then using the two-term AGM inequality, we obtain

$$a_1^{\alpha_1} a_2^{\alpha_2} \ldots a_n^{\alpha_n} \leq \alpha a + \beta b.$$

We now simply apply the $n-1$ term AGM to αa to obtain the desired result.

Similarly we could write down an n term harmonic-geometric mean inequality.

Theorem 5.2.4 Discrete Hölder inequality   Let $p, q > 0$ and $\frac{1}{p} + \frac{1}{q} = 1$. Let $a_1, \ldots a_n, b_1, \ldots, b_n > 0$ be real numbers. Then

$$\sum_{j=1}^{n} a_j b_j \leq \left(\sum_{j=1}^{n} a_j^p\right)^{\frac{1}{p}} \left(\sum_{k=1}^{n} b_k^q\right)^{\frac{1}{q}}.$$

Proof of discrete Hölder inequality

By AGM with $\alpha = \frac{1}{p}$ and $\beta = \frac{1}{q}$, we get

$$ab \leq \frac{a^p}{p} + \frac{b^q}{q}.$$

Applying this to each term in the sum, we get

$$\sum_{j=1}^{n} a_j b_j \leq \frac{1}{p} \sum_{j=1}^{n} a_j^p + \frac{1}{q} \sum_{k=1}^{n} b_k^q.$$

Unfortunately, the right-hand side is always larger than what we want by AGM. However, Hölder's inequality doesn't change if we multiply all a's by a given positive constant and all b's by a given positive constant. So we may restrict to the case that $\sum_{j=1}^{n} a_j^p$ and $\sum_{k=1}^{n} b_k^q$ are both equal to 1. In that case, the right-hand side is exactly 1, which is what we want.

We can also obtain an integral version.

Theorem 5.2.5 Hölder's inequality

Let $p,q > 0$ and $\frac{1}{p} + \frac{1}{q} = 1$. Let f,g be nonnegative integrable functions on an interval $[a,b]$. Then

$$\int_a^b f(x)g(x)dx$$

$$\leq \left(\int_a^b f(x)^p dx\right)^{\frac{1}{p}} \left(\int_a^b g(x)^q dx\right)^{\frac{1}{q}}.$$

To prove this, we just apply the discrete Hölder's inequality to Riemann sums.

We can apply Hölder's inequality to estimate means. To wit, with f nonnegative and integrable and p,q as above:

$$\frac{1}{b-a} \int_a^b f(x)dx$$

$$= \frac{1}{b-a} \int_a^b f(x) \cdot 1 dx$$

$$\leq \left(\frac{1}{b-a} \int_a^b 1^q dx\right)^{\frac{1}{q}} \left(\frac{1}{b-a} \int_a^b f(x)^p dx\right)^{\frac{1}{p}}$$

$$= \left(\frac{1}{b-a} \int_a^b f(x)^p dx\right)^{\frac{1}{p}}$$

This inequality says that the pth root of the mean pth power of f is greater than or equal to the mean of f as long as $p > 1$. Here is slightly more general formulation.

Theorem 5.2.6 Jensen's Inequality

Let g be a convex function and f a nonnegative integrable function. Then

$$g\left(\frac{1}{b-a} \int_a^b f(x) dx\right) \leq \frac{1}{b-a} \int_a^b g(f(x)) dx.$$

We can build up a proof of this starting from sums of two terms, generalizing to sums of n terms by induction, and then ultimately generalizing to integrals by applying the n-term version to Riemann sums. The two-term version,

$$g(\alpha a + \beta b) \leq \alpha g(a) + \beta g(b),$$

is self-evidently the definition of convexity of g. Thus we have come full circle. We can think of Hölder's inequality as being true because the function x^p is convex.

Exercises for Section 5.2

1. Use the concavity of the log function to prove the generalized arithmetic geometric mean inequality: namely if $\alpha, \beta > 0$ and $\alpha + \beta = 1$, then if $a, b > 0$

$$a^\alpha b^\beta \leq \alpha a + \beta b.$$

Flipped recitation, inequalities

In this section, we used convexity to develop some inequalities. Consider the sum

$$(1)(6) + (2)(7) + 3(8) + 4(9) + 5(10) = 130$$

Bound the sum above using the discrete Hölder inequality with the following exponents.

Problem 1

$$p = 2, q = 2$$

Problem 2

$$p = 3, q = \frac{3}{2}$$

Problem 3

$$p = 4, q = \frac{4}{3}$$

◇ **5.3   Economics**

This section will concern economics.

In Section 5.1, we saw that if f,g are functions continuous on $[0,\infty)$ which are twice continuously differentiable on $(0,\infty)$ and satisfy $f''(x),g''(x) < 0$ for all $x \in (0,\infty)$ and also satisfy

$$\lim_{x \longrightarrow 0} f'(x) = \lim_{x \longrightarrow 0} g'(x) = \infty,$$

then the function

$$F_t(x) = f(x) + g(t - x)$$

has a unique maximum. We discussed that this maximization may be thought of as an optimal allocation for a resource for which t units are available and which has two uses whose values are given by f and g.

The current section can be viewed as a very basic introduction to a field called by its practitioners, "modern macroeconomics" which consists entirely in the study of such optimization problems.

A more precise description of macroeconomics is that it is the field of study that concentrates on the behavior of the economy as a whole. Macroeconomics is not well represented at Caltech. (Perhaps one reason is that the reputation of macroeconomics as a subfield of economics is that it is one of the least mathematically rigorous subfields.) However, it is an extremely important subfield, at least in terms of its impact on society. The Federal Reserve Bank, which controls the money supply of the United States (So do many other central banks around the world.) models the economy almost entirely based on theoretical ideas from modern macroeconomics. (It uses statistical input, as well, of course.) Here we'll see a hint of how that works.

Before we can explain how a macroeconomic model works, we have to explain what functions economists are maximizing. To do this, we have to explain the notion of a *utility function*.

Roughly speaking, a utility function $u(x)$ is a function of x, which is a quantity of a good or amount of money. The value of $u(x)$ tells us how happy someone is to have that quantity of the good or that amount of money. You might ask why we need such a function. Why shouldn't I imagine that I am a million times happier having a million dollars than having one dollar? Why not just use x?

Economists often explain that the utility function is a way to avoid gambling paradoxes. Here is a short aside on gambling. A fair coin is one that if you flip it, the probability that it will land on heads is $\frac{1}{2}$ and the probability that it will land on tails is $\frac{1}{2}$. If we play a game of chance where I flip a coin and I give you x dollars when the coin lands on heads and y dollars when the coin lands on tails, we say the "fair price" for playing this game is $\frac{x+y}{2}$ dollars.

Now let's describe a slightly more complicated game of chance. I flip a coin. If it lands on heads, I give you one dollar. If it lands on tails, I flip again. Now if it lands on heads, I give you two dollars. If tails, we flip again. If heads, I give you four dollars, if tails we flip again. And so on. The game ends the first time I flip heads. I pay 2^{j-1} dollars if this happens on the jth try.

What is the fair price of this game? There is $\frac{1}{2}$ probability of heads on the first flip. So this contributes $\frac{1}{2}1 = \frac{1}{2}$ dollars to the fair price. There is $\frac{1}{4}$ probability of getting to the second flip with heads. This contributes $\frac{1}{4}2 = \frac{1}{2}$ dollars to the fair price. There is $\frac{1}{2^j}$ probability of getting to the jth flip and getting heads. This contributes $\frac{1}{2^j}2^{j-1} = \frac{1}{2}$ to the fair price. Each j contributes $\frac{1}{2}$ to the fair price. So the fair price is ∞. No one would pay this price, however.

This creates a problem for economists. They have to explain the behavior of people in the real world. If such people won't pay ∞ for this game, they aren't using the fair price model. Economists explain this by saying that people have a concave utility function. (They just don't like large amounts of money that much.) Really, they are calculating the fair utility they are giving for the utility they might expect. Incidentally, if you ask the same economists why people actually play lotteries, they are likely to say, "Those people are just stupid!"

Now we introduce the simplest version of a modern macroeconomic model. This is sometimes called the neoclassical growth model.

The idea is that people make forecasts about the future and

that they are trying to optimize their happiness by taking the future into account. This will be an optimization problem. In our economy, there will be one kind of good. You might think of seeds. You can eat them or you can plant them. You have a utility function u of x. The number $u(x)$ represents how happy you are when you eat x seeds. The utility function u is a nice function. It is defined and increasing on $(0,\infty)$, it is concave, and

$$\lim_{x \to 0} u'(x) = \infty.$$

The rules of the economy are that you are the only person in the economy. Time is divided into discrete periods. (Think harvests.) In the jth period, you might eat x_j. There is a number $0 < \beta < 1$ called your discounting factor, so that your happiness is given by

$$H(x_0, x_1, \ldots, x_j, \ldots) = u(x_0) + \beta u(x_1) + \beta^2 u(x_2) + \ldots.$$

At time period zero, you start with k_0 seed. You eat x_0 and plant $k_0 - x_0$. The seed you plant goes into a Cobb-Douglas machine as capital. Like the little red hen from the story, you are happy to work to get the harvest. So labor is 1. And $k_1 = (k_0 - x_0)^\alpha$ with $0 < \alpha < 1$. In general, at time period j, you have k_j seed, and you choose $0 < x_j < k_j$ and you get

$$k_{j+1} = (k_j - x_j)^\alpha.$$

Your problem, starting at k_0 is to play this game to optimize $H(x_0, x_1, \ldots, x_j, \ldots)$. How do we do it? It looks hard because it is an optimization problem with infinitely many variables.

The key is to notice that

$$H(x_0, x_1, \ldots x_j, \ldots) = u(x_0) + \beta H(x_1, \ldots, x_j, \ldots).$$

Let $V(k_0)$ be the solution to the game, the optimal value you can get from k_0. Then $V(k_0)$ is the maximum of

$$u(x_0) + \beta V((k_0 - x_0)^\alpha),$$

where x_0 lies between 0 and k_0. This is a resource allocation problem like in Section 5.1, provided that V is a concave function with a derivative going to infinity at 0. However, our reasoning is circular so far. We can only get to this allocation problem by assuming our problem is already solved.

There are various ways of converging to a solution, though. Suppose you know you are only going to live for two time periods. Then, in the last time period, you should eat all your seed. So you are optimizing

$$u(x_0) + \beta u((k_0 - x_0)^\alpha),$$

which you can do. Call the optimum $V_1(k_0)$. Next imagine you will live for three periods. You should optimize

$$u(x_0) + \beta V_1((k_0 - x_0)^\alpha).$$

Call the optimum $V_2(k_0)$. Proceed likewise for arbitrarily many lifetimes and just let

$$V(k_0) = \lim_{j \to \infty} V_j(k_0).$$

Basically the limit will converge, because the difference comes from consumption multiplied by high powers of β which are getting quite small. You should ask how we can prove concavity of the V_j's so we can continue this process. One of the exercises in this section addresses that.

Robert Lucas, who founded modern macroeconomics, got his Nobel Prize for showing that a number of somewhat fancier models can be solved in the same way. The single agent, which I've described as "you", who is the only consumer in this model, plays the role of a representative agent. We figure all consumers are about the same, and we determine how they will behave based on how the representative one behaves. It is possible to think that a macroeconomy consists of a lot of consumers who are different from one another? Can one extend this theory to them? To do this with complete mathematical rigor is a major open problem ...

Exercises for Section 5.3

1. Let $f(K,L)$ be a function of nonnegative K and L with the scaling property

$$f(tK,tL) = tf(K,L),$$

for t, a positive real. Define the single variable functions

$$f_K(L) = g_L(K) = f(K,L).$$

Suppose that each of f_K and g_L is increasing and $f_L(0) = f_K(0) = 0$. Suppose that each of f_K and g_L is continuously differentiable on the positive real numbers. Show that

$$f(K,L) = Kg'_L(K) + Lf'_K(L)$$

for any positive K and L.

2. Let f and g be continuous functions on $[0,\infty)$. Suppose both f and g are twice continuously differentiable on $(0,\infty)$ and that both are concave. Suppose that for each value of t, the function

$$W(x) = f(x) + g(t - x)$$

has a unique maximum attained at $x(t)$ and suppose the function $x(t)$ is twice continuously differentiable. Show that

$$V(t) = f(x(t)) + g(t - x(t))$$

is concave.

Flipped recitation, neo-classical growth

In this section, we introduced the macroeconomic neoclassical growth model for agents of various life spans. Consider the agent who lives for two time periods. Calculate the function $V_2(k)$ for the following choices of u and α.

Problem 1

$$u(x) = \log x, \quad \alpha = \frac{1}{3}$$

Problem 2

$$u(x) = \log x, \quad \alpha = \frac{1}{2}$$

Problem 3

$$u(x) = \sqrt{x}, \quad \alpha = \frac{1}{3}$$

Chapter 6

TRIGONOMETRY, COMPLEX NUMBERS, AND POWER SERIES

◇ 6.1 Trigonometric functions by arclength

We begin by defining the arclength of the graph of a differentiable function $y = f(x)$ between $x = a$ and $x = b$. We motivate our definition by calculating the length of the part of the line $y = mx$ between $x = a$ and $x = b$. This is the hypotenuse of a right triangle whose legs have lengths $b - a$ and $m(b-a)$ respectively. By the Pythagorean theorem, the length of the hypotenuse is given by $(b - a)\sqrt{1 + m^2}$. This motivates the definition of arclength that follows. (One way of thinking about it is that the arclength of a curve is a limit of lengths of piecewise-linear approximations to the curves.)

Arclength   Let f be a differentiable function on the interval $[a,b]$. The *arclength* of the graph of f is

$$\int_a^b \sqrt{1 + (f'(x))^2}\,dx.$$

Let us immediately apply this definition to our favorite curve from plane geometry: the unit circle. The part of the unit circle in the upper half plane is the curve

$$y = \sqrt{1 - x^2}.$$

With $f(x) = \sqrt{1-x^2}$, we calculate

$$f'(x) = \frac{-x}{\sqrt{1-x^2}}.$$

so that

$$\sqrt{1 + (f'(x))^2} = \frac{1}{\sqrt{1-x^2}}.$$

Thus the arclength A of the circle between $x = 0$ and $x = a$ (actually the negative of the arclength if a is negative) is given by

$$A = \int_0^a \frac{dx}{\sqrt{1-x^2}}.$$

We have no especially nice way of computing this integral, so we just give it a name. We say

$$\arcsin a = \int_0^a \frac{dx}{\sqrt{1-x^2}}.$$

This entirely corresponds to our intuition from plane geometry. When studying geometry, we keep going on about the arclengths of parts of circles, even before we can define what arclength actually means. We associate arcs on circles with the angles that subtend them, and the arc we are describing corresponds to the angle whose sine is a. The reason that the inverse to the sine function has such an odd name is that it is computing the length of an arc. For our purposes, we look at things a little differently. We have no way of describing the function $\sin x$ without its inverse, however, because we don't know what the measure of an angle means without the notion of arclength. However clearly arcsin is increasing as a goes from -1 to 1 (and it is an odd function). Thus it has an inverse. We define sin to be the inverse of arcsin. We have not yet named the domain of sin. We make the definition that

$$\frac{\pi}{2} = \int_0^1 \frac{dx}{\sqrt{1-x^2}}.$$

This is really the usual definition for π. It is the arclength of the unit semicircle. We have defined $\sin x$ on the interval $[\frac{-\pi}{2}, \frac{\pi}{2}]$. On the same interval, we may define $\cos x$ by

$$\cos x = \sqrt{1 - \sin^2 x}.$$

It is not hard to see that for $x \in [0, \frac{\pi}{2}]$, we have that

$$\cos(\frac{\pi}{2} - x) = \sin(x),$$

because this is just the symmetry between x and y in the definition of the unit circle.

It is definitely interesting to extend sin and cos to be defined on the whole real line. We are already in a position to do this by symmetry as well, but for the moment we refrain. We will have a much clearer way of defining this extension later when we introduce complex numbers.

But, for now, as long as we stay in the interval $[-\frac{\pi}{2}, \frac{\pi}{2}]$, we are in a position to obtain all the basic facts of calculus for trigonometric functions. Thus, for instance,

$$x = \arcsin(sinx).$$

Differentiating in x, we get

$$1 = \frac{1}{\sqrt{1 - \sin^2 x}}(\frac{d}{dx} \sin x),$$

and solving for the second factor, we obtain the famous formula that

$$\frac{d}{dx} \sin x = \cos x.$$

Applying the symmetry

$$\cos(\frac{\pi}{2} - x) = \sin(x),$$

we immediately obtain that

$$\frac{d}{dx} \cos x = -\sin x.$$

Using these two results, we can easily build up all the famous formulae in the calculus of trigonometric functions.

For instance, we define $\sec x = \frac{1}{\cos x}$ and $\tan x = \frac{\sin x}{\cos x}$. We readily use the quotient rule to calculate

$$\frac{d}{dx} \sec x = \sec x \tan x$$

and

$$\frac{d}{dx}\tan x = \sec^2 x.$$

Then we are free to observe

$$\sec x$$
$$= \frac{\sec x(\sec x + \tan x)}{\sec x + \tan x}$$
$$= \frac{\frac{d}{dx}(\sec x + \tan x)}{\sec x + \tan x}$$
$$= \frac{d}{dx}(\log(\sec x + \tan x)).$$

In short, all the identities of calculus just come to life.

The final thing we bring up here is the dual role of π. Perhaps we all remember π as the arclength of the unit semicircle, but we might also remember it as the area of the unit circle. The first can be a definition, but then the second should be a consequence. Here is how to see it:

We calculate

$$\int_0^1 \sqrt{1 - x^2}dx.$$

This is just the area of one quarter of the unit circle. We will do this using the (quite natural) trigonometric substitution $x = \sin u$. (Wasn't x already the sine of something?) We obtain

$$dx = \cos u\, du.$$

The integral now runs from 0 to $\frac{\pi}{2}$ and becomes

$$\int_0^{\frac{\pi}{2}} \cos^2 u\, du.$$

We calculate this integral without any fancy double-angle identities. We just use again the symmetry

$$\cos(\frac{\pi}{2} - u) = \sin(u),$$

to obtain

$$\int_0^{\frac{\pi}{2}} \cos^2 u\, du = \int_0^{\frac{\pi}{2}} \sin^2 u\, du.$$

Thus

$$\int_0^{\frac{\pi}{2}} \cos^2 u \, du = \frac{1}{2} \int_0^{\frac{\pi}{2}} (\cos^2 u + \sin^2 u) du,$$

and since it is easy to integrate 1, we get

$$\int_0^1 \sqrt{1 - x^2} dx = \frac{\pi}{4}.$$

How is this related to the usual Euclidean proof of the same fact?

1. Prove that π is finite. Hint: You're being asked to show that the improper integral $\int_0^1 \frac{dx}{\sqrt{1-x^2}} = \lim_{a \to 1} \int_0^a \frac{dx}{\sqrt{1-x^2}}$ is finite. The integral has to be considered improper because the integrand goes to ∞ as $x \longrightarrow 1$.) Hint: It might help to use the substitution $y = \sqrt{1 - x^2}$ for part of the integral.

2. Consider the plane curve given by $y = \int_{\frac{\pi}{6}}^x \sqrt{\sec^2 t - 1} dt$ as x runs from $\frac{\pi}{6}$ to $\frac{\pi}{3}$. Calculate its arclength. Hint: Just apply the arclength formula. If you come to an integral whose antiderivative you're having trouble finding, reread the section.

3. Let $f(x)$ be a function which is once continuously differentiable on $[0,1]$. Then the formula for the arclength of its graph is

$$\int_0^1 \sqrt{1 + f'(x)^2} dx.$$

Here is an alternate definiton for the arclength. Let $l_{j,N}$ be the length of the line segment between $(\frac{j-1}{N}, f(\frac{j-1}{N}))$ and $(\frac{j}{N}, f(\frac{j}{N}))$. Let

$$S_N = \sum_{j=1}^N l_{j,N}.$$

Show that

$$\lim_{N \to \infty} S_N = \int_0^1 \sqrt{1 + f'(x)^2} dx.$$

Hint: Can you show that S_N is a Riemann sum for the right hand side?

◇ 6.2   Complex numbers

Here we're going to introduce the system of complex numbers. The main motivation for doing this is to establish a somewhat more invariant notion of angle than we have already. Let's recall a little about how angles work in the Cartesian plane.

We begin by briefly reviewing analytic geometry in two dimensions. Points in the Cartesian plane are given by pairs of numbers (x,y). Usually when we think of points, we think of them as fixed positions. (Points aren't something you add and the choice of origin is arbitrary.) The set of these points is sometimes referred to as the affine plane. Within this plane, we also have the concept of vector. A vector is often drawn as a line segment with an arrow at the end. It is easy to confuse points and vectors since vectors are also given by ordered pairs, but in fact a vector is the difference of two points in the affine plane. (A change of coordinates could change the origin to some other point, but it couldn't change the zero vector to a vector with magnitude.) It is between vectors that we measure angles.

If $\vec{a} = (a_1, a_2)$, we define the magnitude of $\vec{a}$, written $|\vec{a}|$, by $\sqrt{a_1^2 + a_2^2}$, as suggested by the Pythagorean theorem. Given another vector $\vec{b} = (b_1, b_2)$, we would like to define the angle between $\vec{a}$ and $\vec{b}$. We define the dot product

$$\vec{a} \cdot \vec{b} = a_1 b_1 + a_2 b_2.$$

A quick calculation shows that

$$|\vec{a} - \vec{b}|^2 = |\vec{a}|^2 + |\vec{b}|^2 - 2|\vec{a} \cdot \vec{b}|.$$

Therefore, inspired by the law of cosines, we can start to define the angle θ between $\vec{a}$ and $\vec{b}$ by

$$\vec{a} \cdot \vec{b} = |\vec{a}||\vec{b}| \cos \theta.$$

Note that this only defines the angle θ up to its sign. The angle between $\vec{a}$ and $\vec{b}$ is indistinguishable from the angle between $\vec{b}$ and $\vec{a}$ if all we know is the dot product and the magnitudes.

There is another product we can define between two-dimensional vectors which is the cross product:

$$\vec{a} \times \vec{b} = a_1 b_2 - b_1 a_2.$$

We readily observe that

$$|\vec{a} \times \vec{b}|^2 + |\vec{a} \cdot \vec{b}|^2 = |\vec{a}|^2 |\vec{b}|^2.$$

This leads us to

$$\vec{a} \times \vec{b} = |\vec{a}||\vec{b}| \sin\theta,$$

which gives a choice of sign for the angle θ.

We now introduce the complex numbers which give us a way of formalizing a two-dimensional vector as a single number, and defining the multiplication of these numbers in a way that involves both the dot product and cross product of vectors.

Let i to be a formal square root of -1. Of course, the number -1 has no square root which is a real number. i is just a symbol, but we will define multiplication using $i^2 = -1$. A complex number is a number of the form

$$a = a_1 + i a_2,$$

where a_1 and a_2 are real numbers. We write

$$Re(a) = a_1$$

and

$$Im(a) = a_2.$$

We can define addition and subtraction of complex numbers. If

$$b = b_1 + i b_2,$$

then we define

$$a + b = (a_1 + b_1) + i(a_2 + b_2)$$

and

$$a - b = (a_1 - b_1) + i(a_2 - b_2).$$

The addition and subtraction of complex numbers, of course, exactly agree with addition and subtraction of vectors. The fun begins when we define multiplication so that the distributive law holds.

$$ab = a_1 b_1 - a_2 b_2 + i(a_1 b_2 + a_2 b_1).$$

We pause for a quick remark. There is something arbitrary about the choice of i. Certainly i is a square root of -1. But so is $-i$. Replacing i by $-i$ changes nothing about our number system. We name this operation *complex conjugation*. That is, if

$$a = a_1 + i a_2,$$

then the complex conjugate of a is

$$\bar{a} = a_1 - i a_2.$$

Once we have the operation of complex conjugation, we can begin to understand the meaning of complex multiplication. Namely, to the complex number a is associated the vector

$$\vec{a} = (a_1, a_2).$$

Similarly to the complex conjugate of b is associated the vector

$$\vec{b} = (b_1, -b_2).$$

Then

$$ab = \vec{a} \cdot \vec{b} + i\vec{a} \times \vec{b}.$$

To every complex number is associated a magnitude

$$|a| = \sqrt{a_1^2 + a_2^2}.$$

Note that complex conjugation doesn't change this:

$$|a| = |\bar{a}|.$$

To each complex number a is also associated its direction, which we temporarily denote as $\theta(a)$, which is the angle θ that a makes with the x-axis. Complex conjugation reflects complex numbers across the x-axis so

$$\theta(\bar{a}) = -\theta(a).$$

From our description of the multiplication of complex numbers in terms of vectors, we see that

$$ab = |a||b|\cos(\theta(a) + \theta(b)) + i|a||b|\sin(\theta(a) + \theta(b)).$$

Thus

$$|ab| = |a||b|$$

and

$$\theta(ab) = \theta(a) + \theta(b).$$

This gives a geometrical interpretation to multiplication by a complex number a. It stretches the plane by the magnitude of a and rotates the plane by the angle $\theta(a)$. Note that this always gives us

$$a\bar{a} = |a|^2.$$

This also gives us a way to divide complex numbers,

$$\frac{1}{b} = \frac{\bar{b}}{|b|^2},$$

so that

$$\frac{a}{b} = \frac{a\bar{b}}{|b|^2}.$$

There is no notion of one complex number being bigger than another, so we don't have least upper bounds of sets of complex numbers. But it is easy enough to define limits. If $\{a_n\}$ is a sequence of complex numbers, we say that

$$\lim_{n \to \infty} a_n = a$$

if for every real $\epsilon > 0$, there exists $N > 0$ so that if $n > N$, we have

$$|a - a_n| < \epsilon.$$

It is easy to prove that magnitude of complex numbers satisfies the triangle inequality.

In the same way, we can define limits for complex valued functions. A power series

$$\sum_n a_n z^n$$

has the same radius of convergence R as the real power series

$$\sum_n |a_n| x^n$$

and converges absolutely for every z with $|z| < R$.

We can complete our picture of the geometry of complex multiplication by considering

$$e^z = \sum_{n=0}^{\infty} \frac{z^n}{n!}.$$

This power series converges for all complex z since its radius of convergence is infinite. We restrict our attention to the function

$$f(\theta) = e^{i\theta},$$

with θ real. What is $|e^{i\theta}|$? We calculate

$$|e^{i\theta}|^2 = e^{i\theta} \overline{e^{i\theta}} = e^{i\theta} e^{-i\theta} = 1.$$

(Verifying the identity $e^{z+w} = e^z e^w$ is an exercise for this section.) Thus as θ varies along the real line, we see that $e^{i\theta}$ traces out the unit circle. How fast (and in which direction) does it trace it? We get the answer by differentiating $f(\theta)$ as a function of θ. We calculate

$$\frac{d}{d\theta} f(\theta) = i e^{i\theta}.$$

In particular, the rate of change of $f(\theta)$ has magnitude 1 and is perpendicular to the position of $f(\theta)$. We see then that f traces the circle by arclength. That is, θ represents arclength traveled on the circle and from this, we obtain Euler's famous formula

$$e^{i\theta} = \cos\theta + i\sin\theta.$$

By plugging $i\theta$ for z into the definition of e^z and extracting real and imaginary parts, we obtain Taylor series for sin and cos by

$$\sin\theta = \theta - \frac{\theta^3}{3!} + \frac{\theta^5}{5!} + \dots$$

and

$$\cos\theta = 1 - \frac{\theta^2}{2!} + \frac{\theta^4}{4!} + \dots.$$

Exercises for Section 6.2

1. Show that sum of the squares of the lengths of the diagonals of a parallelogram is equal to the sum of the squares of lengths of its sides. Hint: Rewrite the statement in terms of vectors.

2. Prove carefully the identity $e^{x+y} = e^x e^y$ using the power series for e^x. Hint: You can apply the power series to x and y and multiply. Consider all the terms where the number of factors of x and the number of factors of y add to k. Write out what these are and compare them with the binomial theorem.

◇ 6.3   Power series as functions

Recall that a power series is an expression,

$$f(z) = \sum_{j=0}^{\infty} a_j z^j.$$

Here the sequence $\{a_j\}$ may be a sequence of real numbers or of complex numbers, and z may take values in the real numbers or complex numbers.

For any power series there is a nonnegative number R, possibly 0 or ∞, called the radius of convergence of the power series, so that when $|z| < R$, the power series converges absolutely and the power series diverges with $|z| > R$. We can say more specific things. Namely, for any $R' < R$ for any $1 > \rho > \frac{R'}{R}$, there is K depending only on ρ and R' so that when $|z| < R'$, we have

$$|a_j z^j| \le K \rho^j.$$

This simply means that everywhere inside the radius of convergence, the power series may be compared to a convergent geometric series.

When we are interested in restricting the power series to the reals, we might emphasize this by using the variable x and writing

$$f(x) = \sum_{j=0}^{\infty} a_j x^j.$$

For $|x| < R$, the radius of convergence, we may view f as a function of a real variable just like any we've seen so far. We might guess that the derivative of this function is the power series obtained by formally differentiating the original power series.

$$f^{[\prime]}(x) = \sum_{j=1}^{\infty} j a_j x^{j-1}.$$

Note that for $x < R$, the power series $f^{[\prime]}(x)$ converges absolutely. This is clearly true when $x = 0$. For any other x, we compare the jth term of the series to $K\rho^j$. Then the $j-1$st term of $f^{[l]}(x)$ is controlled by $\frac{Kj\rho^j}{x}$. And since

$$\sum_{j=1}^{\infty} j\rho^j$$

converges absolutely, so does $f^{[l]}(x)$.

Theorem 6.3.1

Let

$$f(x)) = \sum_{j=0}^{\infty} a_j x^j$$

and

$$f^{[l]}(x) \sum_{j=1}^{\infty} j a_j x^{j-1}.$$

Let R be the radius of convergence for $f(x)$. Let $|x| < R$. Then f is differentiable at x and

$$f'(x) = f^{[l]}(x).$$

Sketch of Proof of Theorem 6.3.1

Let $|x| < R$. Then there is some R' with $|x| < R' < R$. Because differentiation is local, we can work entirely in the disk $|x| < R'$. There exists K and ρ so that

$$|a_j||y^j| < K\rho^j.$$

Now we set out to calculate

$$\lim_{h \longrightarrow 0} \frac{f(x+h) - f(x)}{h} =$$

$$\frac{\sum_{j=0}^{\infty} a_j (x+h)^j - \sum_{j=0}^{\infty} a_j x^j}{h}.$$

Sketch of Proof of Theorem 6.3.1 cont.

As long as both $|x| < R$ and $|x| + |h| < R$, not only do both sums converge absolutely but we can expand each term $(x+h)^j$ in its binomial expansion and still have absolute convergence, which means we can do the sum in any order we choose. So we reorder the sum according to powers of h. The 0th powers cancel, the 1st powers give the formal derivative, and we get

$$\lim_{h \to 0} \frac{hf^{[\prime]} + \sum_{k=2}^{\infty} \sum_{j=k}^{\infty} a_j \binom{j}{k} x^{j-k} h^k}{h}.$$

We view x as being fixed and the numerator as a power series in h,

$$\lim_{h \to 0} \frac{hf^{[\prime]}(x) + \sum_{k=2}^{\infty} f_k(x) h^k}{h}.$$

We see that the second term in the numerator is h^2 multiplied by a power series in h with a positive radius of convergence. Thus the second term is $O(h^2)$ for h within the radius of convergence and therefore $o(h)$. Thus the theorem is proved.

Having discovered that we can differentiate power series formally, we see that a lot of calculations with series become much easier. Here are some examples.

Example 1   Calculations with power series

We've known for a long time that when $|x| < 1$, we have

$$\frac{1}{1-x} = 1 + x + x^2 + x^3 + \dots.$$

We can take the derivative of both sides, obtaining

$$\frac{1}{(1-x)^2} = 1 + 2x + 3x^2 + \dots.$$

Of course, we also have

$$\frac{1}{(1-x)^2} = (1 + x + x^2 + \dots)^2,$$

obtaining

$$(1 + x + x^2 + \dots)^2 = 1 + 2x + 3x^2 + \dots.$$

Indeed, we can also integrate the equation

$$\frac{1}{1-x} = 1 + x + x^2 + x^3 + \dots,$$

obtaining

$$-\log(1-x) = x + \frac{x^2}{2} + \frac{x^3}{3} + \dots,$$

using the fact that $\log 1$ is 0.

If we have a lot of faith in the theory of differential equations, we might suppose that e^x is the only solution to the the equation

$$f'(x) = f(x),$$

with

$$f(0) = 1.$$

We then readily see that

$$f(x) = \sum_{j=0}^{\infty} \frac{x^j}{j!}.$$

This is an independent way of deriving the power series for e^x. Plugging in ix, we see that we have power series for sin and cos too.

An alternative way to think about the derivation of e^{ix} is that it is the unique solution to

$$f'(x) = if(x),$$

with

$$f(0) = 1.$$

Of course, $f(x) = \cos x + i \sin x$ solves this too. The differential equation can be interpreted as saying the tangent line is perpendicular to the radius.

Exercises for Section 6.3

1. Let $\sum_{j=1}^{\infty} a_j$ be an absolutely convergent series. Show that the sum of this series is independent of the order in which you add the terms. Hint: Since the series is absolutely convergent, for every $\epsilon > 0$ there exists a number N so that the sum of absolute values $|a_j|$ with $j > N$ is less than ϵ. Now consider some other ordering of the a_j's. There is some M for which $a_1, a_2, \ldots, a_N$ appear before the Mth term of the new ordering.

Chapter 7

COMPLEX ANALYSIS

◇ 7.1 Roots of polynomials

In Chapter 6, we described the complex numbers. Classically, one of the reasons complex numbers were studied is that in the complex numbers all polynomials have roots. We discuss this a bit here.

We derived Euler's formula:

$$e^{i\theta} = \cos\theta + i\sin\theta.$$

Because as θ increases, $e^{i\theta}$ traces the unit circle counterclockwise, every complex number z with $|z| = 1$ can be written as $e^{i\theta}$ for some θ. Because both sin and cos are 2π periodic, we have that

$$e^{i\theta} = e^{i\theta + 2\pi n},$$

for each integer n. Thus every complex number of magnitude 1 is associated with an angle $\theta + 2\pi i n$ up to ambiguity about the choice of n.

Let z be a general complex number. We can write

$$|z| = r,$$

with r a nonnegative real number, and if r differs from 0, we can consider $\frac{z}{|z|}$, which is a complex number with magnitude 1, and write

$$\frac{z}{|z|} = e^{i\theta}.$$

Thus any complex number z may be written as

$$z = re^{i\theta},$$

and this representation of points in the plane is sometimes referred to as the polar coordinate system.

One of the simplest polynomials we can write is

$$x^k - z = 0,$$

with z a complex number. The roots of this polynomial are called the kth roots of z. In particular, if

$$z = re^{i\theta},$$

then one root of the polynomial is given by

$$x = r^{\frac{1}{k}} e^{\frac{i\theta}{k}}.$$

Of course, because θ has some ambiguity, so does the definition of x, and we can write

$$x_j = xe^{\frac{2\pi ij}{k}},$$

with j running over integers $0, \ldots, k-1$. The numbers $e^{\frac{2\pi ij}{k}}$ are referred to as the kth roots of unity (that is, roots of 1).

Thus every complex number has k many kth roots for every natural number k. There are, of course, other polynomials than $x^k - z$. We would like to show that every polynomial with complex coefficients has a complex root. One early approach to the problem of showing that polynomials have roots was to give the roots in terms of kth roots of expressions involving the coefficients of the polynomial.

An example you are probably all familiar with is the quadratic formula. The goal is to find all solutions of

$$ax^2 + bx + c = 0,$$

with $a \neq 0$. It is a simple matter to divide out a. Letting $p = \frac{b}{a}$ and $q = \frac{c}{a}$, we get

$$x^2 + px + q = 0.$$

(Incidentally, this is a polynomial whose roots have product q and sum $-p$.) we can complete the square, introducing $y = x + \frac{p}{2}$ to get a polynomial, the sum of whose roots is zero.

$$y^2 + q - \frac{p^2}{4} = 0.$$

The roots here are just the square roots of $q - \frac{p^2}{4}$. Now we can subtract $\frac{p}{2}$ to recover x from y and plug into our expression for x, the definitions of p and q and we recover the familiar quadratic formula.

A similar approach can be used to get the root of a general cubic. This was first done in Renaissance Italy, probably by Niccolò Tartaglia. The formula bears the name "Cardano's formula" because Tartaglia kept his technique a secret, whereas Cardano bought it from him and published it.

As above, we can divide by the leading coefficient and complete the cube to reduce the general cubic to an equation,

$$t^3 + pt + q = 0.$$

Next, we give ourselves some flexibility by introducing two unknowns, u and v. We let

$$u + v = t,$$

which still gives us one degree of freedom and allows us to choose the product uv. We set

$$3uv + p = 0.$$

Since plugging $u + v$ into the cubic gives us

$$u^3 + v^3 + (3uv + p)(u + v) + q = 0,$$

we get

$$u^3 + v^3 = -q$$

$$u^3 v^3 = -\frac{p^3}{27}.$$

Thus u^3 and v^3 are the roots of the quadratic

$$z^2 + qz - \frac{p^3}{27}.$$

We now take cube roots of the two roots of the quadratic and match them so that $uv = \frac{-p}{3}$. We then use $t = u + v$ to get the root of a cubic.

Unfortunately, in the early nineteenth century, the great mathematician Niels Abel showed that in general, quintics, polynomials in degree 5, cannot be solved just by taking repeated kth roots. Nonetheless, the following holds.

Theorem 7.1.1
Fundamental theorem of algebra

Let $p(z)$ be a polynomial with complex coefficients and degree $k \geq 1$. Then there is a complex number z with $p(z) = 0$.

We sketch a proof.

It suffices to consider a polynomial of the form

$$p(z) = z^k + c_1 z^{k-1} + \cdots + c_k,$$

with $c_k \neq 0$. (If $c_k = 0$, then we already know $z = 0$ is a root.)

Let us consider the complex valued functions

$$f_r(\theta) = p(re^{i\theta})$$

on $[0, 2\pi]$. The ranges of these functions are closed curves in the complex plane. When r is very small, these curves live very close to c_k and do not enclose the origin at all. When r is very large, then the function $f_r(\theta)$ resembles $r^k e^{ik\theta}$. Thus $f_r(\theta)$ lies on a curve relatively close to the circle of radius r^k and in fact winds around the origin k times. If we change r just a little, it has no effect on how many times $f_r(\theta)$ winds around the origin unless the curve crosses the origin as we change r. But this means that p has a root.

◇ 7.2   Analytic functions

In this section, we will explore the calculus of complex-valued functions of a complex number. Already you have frequently met complex-valued functions $f(t)$ of a real variable t. Complex numbers have real and imaginary parts so we can write

$$f(t) = u(t) + iv(t).$$

Then, if the derivative exists,

$$f'(t) = u'(t) + iv'(t),$$

which is consistent with

$$f'(t) = \lim_{h \longrightarrow 0} \frac{f(t+h) - f(t)}{h}.$$

We can take any differentiable functions $u(t)$ and $v(t)$ of one real variable and combine them as $u(t) + iv(t)$ and obtain a complex-valued function of a real variable which is differentiable.

When it comes to functions of a complex variable, something more special is going on. All "elementary functions", those really formulaic functions that we meet in calculus, can be applied to complex numbers. To wit:

$$e^{x+iy} = e^x e^{iy} = e^x \cos y + ie^x \sin y.$$

We will take e^z with z a complex variable as an example throughout this section.

Further,

$$\log(x + iy) = \log(\sqrt{x^2 + y^2}) + \log(\frac{x + iy}{\sqrt{x^2 + y^2}})$$

$$= \log(\sqrt{x^2 + y^2}) + i \arcsin(\frac{y}{\sqrt{x^2 + y^2}}).$$

Even trigonometric functions can be applied to a complex variable z.

$$\sin z = \frac{e^{iz} - e^{-iz}}{2i}.$$

Often sin applied to an imaginary argument is referred to as hyperbolic sin.

We will use

$$f(z) = f(x + iy) = e^{x+iy} = e^x \cos y + ie^x \sin y,$$

as our main example.

A complex-valued function $f(z)$ has real and imaginary parts

$$f(x + iy) = u(x,y) + iv(x,y),$$

each of which is a real function of two real variables. In the case of e^z, we have

$$u(x,y) = e^x \cos y$$

and

$$v(x,y) = e^x \sin y.$$

We will briefly discuss the differentiation theory of real functions of two variables. You can view this as preview of multivariable calculus.

Recall that a function of one variable $f(x)$ is differentiable at x exactly when there is a number $f'(x)$ so that

$$f(x + h) = f(x) + f'(x)h + o(h).$$

Given a function $f(x,y)$ of two variables, we can leave y fixed and view it purely as a function of x and then take the derivative. This derivative $\frac{\partial f}{\partial x}$ is called the partial derivative in x and is defined by

$$f(x + h,y) = f(x,y) + \frac{\partial f}{\partial x}(x,y)h + o(h).$$

Similarly $\frac{\partial f}{\partial y}(x,y)$ is defined by

$$f(x,y + k) = f(x,y) + \frac{\partial f}{\partial y}(x,y)k + o(k).$$

If $\frac{\partial f}{\partial x}$ and $\frac{\partial f}{\partial y}$ are defined and *continuous* at all points near (x,y) then something stronger is true. Namely,

$$f(x + h, y + k)$$

$$= f(x + h, y) + \frac{\partial f}{\partial y}(x + h, y)k + o(k)$$

$$= f(x,y) + \frac{\partial f}{\partial x}(x,y)h + \frac{\partial f}{\partial y}(x + h, y)k + o(h) + o(k)$$

$$= f(x,y) + \frac{\partial f}{\partial x}(x,y)h + \frac{\partial f}{\partial y}(x,y)k + o(1)k + o(h) + o(k)$$

$$= f(x,y) + \frac{\partial f}{\partial x}(x,y)h + \frac{\partial f}{\partial y}(x,y)k + o(h) + o(k).$$

Note that the equality between the third and fourth line used the continuity of $\frac{\partial f}{\partial y}$. The equality between the first and fifth line is usually taken as the definition of differentiability and is the two-variable differential approximation.

Gottfried Wilhelm Leibniz introduced shorthand for differential approximations. The one variable differential approximation for $f(x)$, a differentiable function of one variable, becomes

$$df = \frac{df}{dx}dx.$$

Shorthand for the differential approximation of $f(x,y)$, a differentiable function of two variables, is

$$df = \frac{\partial f}{\partial x}dx + \frac{\partial f}{\partial y}dy.$$

The expression df is called the *differential* of f

Now we are ready to see what is the differential of e^z, our central example.

$$f(x,y) = e^x \cos y + ie^x \sin y.$$

We calculate

$$df = (e^x \cos y + ie^x \sin y)dx + (-e^x \sin y + ie^x \cos y)dy$$
$$= e^x \cos y(dx + idy) + e^x \sin y(idx - dy)$$
$$= (e^x \cos y + ie^x \sin y)(dx + idy)$$
$$= e^z dz.$$

!!! Here we use $z = x + iy$ and $dz = dx + idy$. What we have seen is that with $f(x,y) = e^z$, we have the differential approximation $df = e^z dz$.

This motivates a definition.

Analytic function

> A complex-valued function of a complex variable $f(z)$ which is differentiable at z as a function of two variables is *analytic* at z if $\frac{df}{dz}$ cleans up, that is, if $df(z)$ is a complex multiple of dz.

Analyticity is a very special condition. It is not the case that if we write

$$f(z) = u(x,y) + iv(x,y),$$

for any nice functions u and v that we will end up with an analytic function. Let's see what conditions analyticity puts on u and v.

$$df = du + idv$$
$$= \frac{\partial u}{\partial x}dx + \frac{\partial u}{\partial y}dy + i(\frac{\partial v}{\partial x}dx + \frac{\partial v}{\partial y}dy)$$
$$= [\frac{\partial u}{\partial x} + i\frac{\partial v}{\partial x}]dx + [\frac{\partial u}{\partial y} + i\frac{\partial v}{\partial y}]dy.$$

Now it is clear that $\frac{df}{dz}$ cleans up means, that

$$\frac{\partial u}{\partial y} + i\frac{\partial v}{\partial y} = i(\frac{\partial u}{\partial x} + i\frac{\partial v}{\partial x}).$$

Breaking this equation into real and imaginary parts, we get

$$\frac{\partial u}{\partial y} = -\frac{\partial v}{\partial x}.$$
$$\frac{\partial v}{\partial y} = \frac{\partial u}{\partial x}.$$

These equations, giving the conditions on u and v for $u+iv$ to be analytic, are called the Cauchy-Riemann equations.

There are various alternative interpretations of the condition of analyticity. In particular, $f(z)$ is analytic at z if

$$\lim_{h \longrightarrow 0} \frac{f(z+h) - f(z)}{h}$$

exists as a complex limit (that is, with h running over the complex numbers). The reason that this is so much more restrictive than ordinary differentiation, is , again the role played by complex multiplication.

To have a complete picture of analytic functions of a complex variable, it is not enough just to be able to differentiate. We should also be able to integrate. In the next section, we'll worry about when it makes sense to integrate an expression of the form $f(z)dz$ with f an analytic function.

For a moment, we leave that as a mystery. One clue is that if we write

$$f = u + iv,$$

then

$$f(z)dz = udx - vdy + i(vdx + udy).$$

A perhaps more satisfying clue is that if F is a complex antiderivative for f, we have that

$$dF = f(z)dz.$$

In the next section, we wil consider when does $f(z)$ analytic have a complex antiderivative?

◇ 7.3 Cauchy's theorem

In the previous section, we introduced the notion of a differential of a function of two variables $f(x,y)$; namely,

$$df = \frac{\partial f}{\partial x}dx + \frac{\partial f}{\partial y}dy.$$

We said that a complex-valued function of a complex variable is analytic if

$$df = f'(z)dz,$$

that is, if the differential of f is a complex multiple of dz. And we ended with the following question: Given an analytic function $f(z)$, when can we find any analytic antiderivative, a function $F(z)$ so that

$$dF = f(z)dz?$$

In thinking about how to answer this, we can first ask the more general question: given a differential form

$$\omega = u(x,y)dx + v(x,y)dy,$$

when can we find $F(x,y)$ so that

$$dF = \omega?$$

(We'll assume that $u(x,y)$ and $v(x,y)$ are continuous and everywhere differentiable.)

Thus we are trying to solve the equations

$$\frac{\partial F}{\partial x} = u$$
$$\frac{\partial F}{\partial y} = v$$

since, of course,

$$dF = \frac{\partial F}{\partial x}dx + \frac{\partial F}{\partial y}dy.$$

These are two equations for only one unknown function so we should not expect it always to be possible to solve them.

Fact

> Let F be a function of two variables whose first partial derivatives have continuous first partial derivatives. Then
>
> $$\frac{\partial}{\partial y}\left(\frac{\partial F}{\partial x}\right) = \frac{\partial}{\partial x}\left(\frac{\partial F}{\partial y}\right).$$
>
> We refer to the quantity on both sides of this equation as $\frac{\partial^2 F}{\partial x \partial y}$, the mixed second partial derivative of F.

How would we go about proving the fact? We note that the left-hand side of the equality is given by

$$\frac{\partial}{\partial y}\left(\frac{\partial F}{\partial x}\right) =$$

$$\lim_{k \longrightarrow 0} \lim_{h \longrightarrow 0} \frac{F(x+h,y+k) - F(x+h,y) - F(x,y+k) + F(x,y)}{hk}$$

while the right hand side is given by

$$\frac{\partial}{\partial x}\left(\frac{\partial F}{\partial y}\right) =$$

$$\lim_{h \longrightarrow 0} \lim_{k \longrightarrow 0} \frac{F(x+h,y+k) - F(x+h,y) - F(x,y+k) + F(x,y)}{hk}.$$

The only difference between the two is the order in which the limit is taken. A careful ϵ-δ argument shows they are the same. Thus if we have $\frac{\partial F}{\partial x} = u$ and $\frac{\partial F}{\partial y} = v$, then we must have

$$\frac{\partial u}{\partial y} = \frac{\partial v}{\partial x}.$$

We would like to see that the necessary condition is also sufficient. By the single variable fundamental theorem of calculus, we can solve $\frac{\partial F}{\partial x} = u$ by

$$F(x,y) = \int_{x_0}^{x} u(s,y)ds + C_1(y),$$

and similarly we can solve $\frac{\partial F}{\partial y} = v$ by

$$F(x,y) = \int_{y_0}^{y} v(x,t)dt + C_2(x).$$

All this works out if

$$G(x,y) = \int_{x_0}^{x} u(s,y)ds - \int_{y_0}^{y} v(x,t)dt,$$

is the difference between a function of x and a function of y. To verify this, we just need to check that the second mixed partial vanishes. We calculate

$$\frac{\partial G}{\partial x}$$

$$= u(x,y) - \int_{y_0}^{y} \frac{\partial v}{\partial x}(x,t)dt$$

$$= u(x,y) - \int_{y_0}^{y} \frac{\partial u}{\partial y}(x,t)dt$$

$$= u(x,y_0).$$

Taking the derivative in y, we calculate that

$$\frac{\partial^2 G}{\partial x \partial y} = 0,$$

as desired. Thus everything works. We can find the antiderivative F as long as u and v satisfy the consistency condition

$$\frac{\partial u}{\partial y} = \frac{\partial v}{\partial x}$$

and as long as all derivatives are defined and continuous on the rectangle with opposite corners (x_0, y_0) and (x, y).

So how does this work for $f(z)dz$ when

$$f(x + iy) = u(x,y) + iv(x,y)$$

is analytic? We calculate that

$$f(z)dx = (u + iv)(dx + idy) = (u + iv)dx + (-v + iu)dy.$$

We check the consistency condition

$$\frac{\partial u}{\partial y} + i\frac{\partial v}{\partial y} = -\frac{\partial v}{\partial x} + i\frac{\partial u}{\partial x}.$$

These are the same by the Cauchy-Riemann equations and so everything works. We can find $F(z)$ so that

$$dF = f(z)dz.$$

By definition, $F(z)$ is analytic and an anti-derivative of f.

Now how (and on what) can we integrate an expression like dF? We integrate on a curve, a complex-valued function α on a real interval $[a,b]$. We view α geometrically as a curve between $\alpha(a)$ and $\alpha(b)$.

We define

$$\int_\alpha dF = \int_a^b f(\alpha(t))\alpha'(t)dt,$$

where the multiplication, of course, is complex multiplication. We observe that since $dF = f(z)dz$, the integral becomes

$$\int_a^b \frac{d}{dt}F(\alpha(t))dt = F(\alpha(b)) - F(\alpha(a)).$$

This is rather amazing. For analytic functions, integration on a curve is just like one-variable integration of a real function. You don't care about the path of the curve. The integral is the difference of the values of the antiderivative at the endpoints.

Usually, this is stated in terms of closed curves, those where $\alpha(a) = \alpha(b)$. We arrive at

Theorem 7.3.1 Cauchy's theorem

Let f be analytic with continuous derivative on a rectangle R. Let α be a closed curve lying in the rectangle R. Then

$$\int_\alpha f(z)dz = 0.$$

The catch here is the function must be analytic on a rectangle containing the curve for our argument to work. What happens when that isn't the case? We consider a central example.

Suppose we want to integrate $\frac{dz}{z}$. Does $\frac{1}{z}$ have an antiderivative? It should be $\log z$. Recall that

$$\log z = \log|z| + i\theta,$$

where

$$\theta = \arcsin \frac{y}{\sqrt{x^2 + y^2}}.$$

The problem is that θ doesn't have a continuous value on any rectangle containing the origin. (Alternatively, the problem is that $\frac{1}{z}$ is not analytic at 0.) So let's integrate $\frac{dz}{z}$ on the closed curve

$$\alpha(t) = e^{2\pi i t},$$

defined on $[0,1]$. We get

$$\int_\alpha \frac{dz}{z} = \int_0^1 \frac{2\pi i e^{2\pi i t} dt}{e^{2\pi i t}} = \int_0^1 2\pi i \, dt = 2\pi i.$$

Why? Because θ changes by 2π as we go around the circle. The integral measures how many times the curve winds around the singularity at 0. In the next section, we will use this idea to discover one more nice property of analytic functions.

◇ 7.4 Taylor series of analytic functions

In this section, our goal is to show all analytic functions are given locally by convergent power series.

From the previous section we have:

If α is a closed curve contained in a rectangle R where f is analytic, then

$$\int_\alpha f(z)dz = 0.$$

We also saw that if α was a circle centered at 0

$$\int_\alpha \frac{dz}{z} = 2\pi i.$$

In fact, as long as the circle surrounds 0,

$$\int_\alpha \frac{dz}{z} = 2\pi i.$$

because

$$d\log z = \frac{dz}{z}.$$

Now we want to combine these facts. We'll begin with a function analytic in a small rectangle R containing 0 in its interior. We'll let α be a circle of some radius r centered at 0 but with r small enough so that the circle is contained in the rectangle R. Finally, for convenience we'll also assume that $f'(z)$ is analytic in R. (But we're only assuming that we have two derivatives of f.)

Fact

$$0 = \frac{1}{2\pi i}\int_\alpha \frac{(f(z) - f(0))dz}{z}.$$

If the fact is true, then we get

$$f(0) = \frac{1}{2\pi i}\int_\alpha \frac{f(0)dz}{z} = \frac{1}{2\pi i}\int_\alpha \frac{f(z)dz}{z}.$$

To confirm the fact, we need that $\frac{f(z)-f(0)}{z}$ is analytic on R. Clearly this holds at every point but $z = 0$. Moreover, the function is continuous at 0, limiting to $f'(0)$. We just need to check if it has a derivative there. We calculate

$$\lim_{h \longrightarrow 0} \frac{\frac{f(h)-f(0)}{h} - f'(0)}{h}$$

$$= \lim_{h \longrightarrow 0} \frac{f(h) - f(0) - hf'(0)}{h^2}$$

$$= \lim_{h \longrightarrow 0} \frac{\frac{h^2}{2}f''(0) + o(h^2)}{h^2}$$

$$= \frac{1}{2}f''(0).$$

So we conclude that

$$f(0) = \int_\alpha \frac{f(z)dz}{z}.$$

Now we apply the same argument for w near 0 (in particular, on the inside of the circle α). We get

$$f(w) = \frac{1}{2\pi i} \int_\alpha \frac{f(z)dz}{z - w}.$$

What's amazing about this formula is that we've moved the entire dependence of the function on w inside this integral. In fact, all of the behavior of f is determined by its values on the circle α.

Now $f(z)$ is continuous on α, so by the extreme value theorem, it is bounded. That is

$$|f(z)| \leq M$$

for some real number M when z is on the circle α.

Moreover we can calculate derivatives by differentiating under the integral sign. (We leave this as an exercise to the reader, but the point is that the dependence on w in the denominator of the integrand is analytic.)

We get

$$f'(w) = \frac{1}{2\pi i} \int_\alpha \frac{f(z)dz}{(z - w)^2}.$$

$$f''(w) = \frac{1}{2\pi i} \int_\alpha \frac{2f(z)dz}{(z-w)^3}.$$

$$\cdots$$

$$f^{(n)}(w) = \frac{1}{2\pi i} \int_\alpha \frac{n!f(z)dz}{(z-w)^{n+1}}.$$

Thus

$$|f^{(n)}(0)| = |\frac{1}{2\pi i} \int_\alpha \frac{n!f(z)dz}{(z)^{n+1}}|$$

$$\leq \frac{1}{2\pi} \int_\alpha |\frac{n!f(z)}{(z)^{n+1}}|dz$$

$$\leq \frac{Mn!}{r^n}.$$

Thus if we write out the Taylor series of f at 0, we get

$$\sum_{n=0}^\infty \frac{f^{(n)}(0)}{n!} z^n.$$

The nth coefficient is bounded in absolute value by $\frac{M}{r^n}$. These are exactly the coefficients for the series for $\frac{M}{1-\frac{z}{r}}$. By comparison, we see the Taylor series has a radius of convergence of at least r.

But how do we know that the Taylor series describes the function? We have one last trick at our disposal. We return to the formula

$$f(w) = \frac{1}{2\pi i} \int_\alpha \frac{f(z)dz}{z-w}.$$

Notice that

$$\frac{1}{z-w} = \frac{1}{z}\frac{1}{1-\frac{z}{w}} = \frac{1}{z}(1 + \frac{z}{w} + (\frac{z}{w})^2 + \ldots).$$

By expanding the integrand in a series in w, we get our convergent Taylor series.

Our conclusion is that all the elementary functions of calculus and indeed all analytic functions have convergent power series. Indeed, all the fancy complex calculus we've been doing with them could have been concluded just from that. (In the nineteenth century, people tried to expand the list of elementary functions by adding in their favorite "special functions" which are defined as power series. But much of this knowledge is now largely forgotten.)

We had to pass to complex numbers to make these arguments. When we dealt with Taylor series for real functions, we could find finite Taylor approximations, but under very reasonable conditions like C^∞, we had no guarantee of a Taylor series, even if the series converged, converged to the value of the function. The complex numbers overshadow the elementary calculus of one variable silently pulling its strings.

With that revelation, we end the book.

Appendix A

APPENDIX

◇ A.1 How to Use This Book

Here is a brief set of suggestions tfor the self-studying crank for how to use this book.

My main suggestion is to read the book. That may sound obvious, but it is not the approach most students take to mathematics books. Often students will just try to do the problems. If they have access to the author of the book, their most common question will be "What do you want me to do?" Part of being a crank is being able to accept the answer "I don't want you to do anything. You can do what you like."

There is more to be said about how to read a math book. It is different from reading a novelwhere the fundamental illusion is that what is happening to the characters is happening to the reader. The reader of a math book, at least if that reader is a crank, should develop the illusion of having invented the subject. The book does list definitions, theorems and proofs, but if you are a crank, you might want to reorganize them in the way that seems most logical to you.

Don't just read: read with paper and pencil at hand. All definitions should be read and understood, of course, but you shouldn't be entirely uncritical of them. "Why is this being defined?" you should ask. "Why does it interest me?" Whenever a theorem (or lemma, or proposition) is being proved, you should start not by reading the proof but by trying to fill it in yourself. Only if you have struggled with the proofs are you likely to fully understand them. Even after you have read the proof, you might still wonder if the proof is natural and whether the same thing can be proved in another way. Don't be afraid to think creatively about the material.

All sections up to the end of Chapter 5 come with two kinds of problems, "flipped recitations" and "exercises." The flipped recitations are intended as routine questions to test basic understanding of the material. In a classroom setting, they are intended to be solved by students in one class session, with some assistance, if necessary, from a teacher, Professor, or teaching assistant.

The exercises often let you develop the material more deeply than is done in the text. They are not exhaustive by any means. Other questions could be asked. Most of the exercises are not too difficult and hints are given to point you in the right direction. The main difficulty in doing them is to make sure everything you are saying is true. Some students think that everything that needs to be said to solve the problems is obvious. But you have to be careful. A thing is obvious if you know why it is true. Unfortunately some students think things are obvious which are false.

If you're reading this book and determined to be a crank, then congratulations! Have fun. You have a chance at becoming a master of the subject.

◇ **A.2   Logic, Axioms, and Problem Solving**

Students at Caltech who take Math 1a spend the preceding summer taking an online course on formal logic called Math 0. In a few paragraph, we can try to touch on some of the highlights. Logic should be intuitive. In some sense, it is ingrained in humans.

Logic is the study of statements, that might be either true or false. Sometimes, we represent tha statement with a single letter to emphasize how it is related to other statements.

Example 1   Suppose that P and Q are each statements.
A new statement   "If P then Q." is a statement

A typical way you might prove that the statement Q is true is to knpw the statement P is true and that the statement "if P then Q" is true. (Such a proof is sometimes called a syllogism.)

Warning: "If P then Q" and "If Q then P" are different statements. Each is called the *converse* of the other. However, "If not Q then not P" is the *contrapositive* of "If P then Q" and is equivalent to it. Very often, we prove statements by proving their contrapositives. This is essentially the same as a proof by contradiction. You assume what you want to prove is not true and you arrive at a contradiction.

Finding contrapositives and proofs by contradiction requires being able to negate statements, that is, being able to construct the statement not P from the statement P. When reading this book, that problem will often arise in the context of *quantifiers*.

Often statements will have parameters. The statement $P(x)$ will be different for different values of the variable x. You might be interested in statements like "for all x, $P(x)$" or "there exists an x so that $P(x)$." The first statement can be disproved by finding an x for which $P(x)$ is false. (This is called finding a counterexample.) The second statement can be proved by finding an x. The negation of "for all x, $P(x)$" is "there exists x, with not $P(x)$" Likewise, the negation of "there exists x, $P(x)$" is "for all x, not $P(x)$."

Most of the statements made in this book will concern numbers. Usually systems of numbers are defined with axioms. We try to largely avoid this. We will assume throughout the book that you know everything about the natural numbers and the rational numbers. This is because the arithmetic you learn in grade school concerns only finite processes. The exception is proof by induction which is emphasized in the first few sections of the book. When it comes to real numbers, we should be more careful.

Often the real numbers are defined by giving a list of their axioms. The real numbers are in fact the only number system that satisfies these axioms.

Axioms for the real numbers For all x,y,z real numbers

- $x + y$ is real

- **addition is commutative** $x+y = y+x$

- **addition is associative** $(x + y) + z = x + (y + z)$

- There is a real number 0 with $0 + x = 0$

- There is a number $-x$ with $x+(-x) = 0$

- The product xy is real.

- **multiplication is commutative** $xy = yx$

- **multiplication is associative** $(xy)z = x(yz)$

Axioms for the real numbers cont.

- There is a real number 1 so that $1x = x$

- For every nonzero x, there is a number $\frac{1}{x}$ with $x\frac{1}{x} = 1$

- **distributive law** $(x + y)z = xy + xz$

- **order** Exactly one of the following is true $x < y, \quad x = y, \quad x > y$

- **transitivity** If $x < y$ and $y < z$ then $x < z$

- If $x < z$ then $x + y < x + z$

- If $x > 0$ and $y > 0$ then $xy > 0$

- **least upper bound property** Every nonempty subset of the real numbers which is bounded above has a least upper bound

We will not use the axioms to construct the real numbers. Instead, we will define them explicitly, and from our definition, we can prove all the axioms. (We do not fully do this because it is a bit tedious.) You should note that the rational numbers satisfy all axioms but the last. When proving these other axioms, we will use the fact that the rationals satisfy them. This is not circular reasoning. You are expected to know everything about the rationals because addition and multiplication of rationals are finitie processes. But the same is not true for general real numbers. By the time you start Chapter 2, you may use the axioms for the reals freely.

Many students claim the exercises in this book are hard. For the most part, I don't believe this. Students can find solving problems difficult when they don't understand what is being asked. (This is one reason it is important to read the section before attempting its exercises. Another reason is that the exercises do not exist for their own sake. They exist to illustrate the ideas of the section.) To attempt an exercise, figure how all the words are defined. Then try to unwind those definitions into a solution.

◇ A.3   Further Reading

You will be sad to learn that not all of mathematics is covered in this book. The purpose of this section is to make some suggestions for further reading.

The modern way of presenting Real Analysis, the foundation of Calculus, may have started with Richard Courant's book "Differential and Integral Calculus". (1934) A somewhat more recent and palatable treatment is Rudin's "Principles of Mathematical Analysis." (3rd Edition 1976). In the 1960's there were at least three Calculus books which took a rigorous approach, as opposed to the intuitive approach of books entitled "Calculus" by authors such as George B. Thomas, James Stewart and their fellow travelers. The rigorous books were by Tom Apostol "Calculus" (1967), Michael Spivak "Calculus" (1967), and Joseph W. Kitchen jr. "Calculus of One Variable" (1968). Of the three, I like Kitchen's book best for his many nice problems about exponential functions.

In the Caltech core (and pseudo-core) curriculum, the calculus course is followed by courses on linear algebra, multivariable calculus, and ordinary differential equations. My favorite linear algebra is Treil's book "Linear Algebra Done Wrong", which is available free online, and has everything one needs to know about vector spaces if one is content to work over the real and complex fields. A beautiful book on multivariable Calculus is Michael Spivak's "Calculus on Manifolds" (1968). A classic rigorous text on differential equations is that of Earl A. Coddington and Norman Levinson (1955).

If you like this book, though, you could just go to the math section of your local university library and just read every book you can get your hands on. What could possibly go wrong?

INDEX